# Wildlife and Roads
## THE ECOLOGICAL IMPACT

An Occasional Publication of the Linnean Society of London

# Wildlife and Roads
## THE ECOLOGICAL IMPACT

*editors*

## Bryan Sherwood
Linnean Society of London, UK

## David Cutler
Royal Botanic Gardens, Kew, UK

## John A. Burton
Consultant

Imperial College Press

*Published by*

Imperial College Press
57 Shelton Street
Covent Garden
London WC2H 9HE

*Distributed by*

World Scientific Publishing Co. Pte. Ltd.
P O Box 128, Farrer Road, Singapore 912805
*USA office:* Suite 1B, 1060 Main Street, River Edge, NJ 07661
*UK office:* 57 Shelton Street, Covent Garden, London WC2H 9HE

**Library of Congress Cataloging-in-Publication Data**
Wildlife and roads : the ecological impact / edited by David Cutler & John Burton.
  p. cm.
  Revised papers originally presented at a conference held at the Linnean Society of
London in 1998.
  Includes bibliographical references.
  ISBN 1-86094-321-7 (alk. paper)
    1. Roads--Environmental aspects--Congresses. I. Cutler, D. F. (David Frederick),
1939- II. Burton, John A.

  QH545.R62 W56 2002
  577.27'2--dc21                                                                      2002024249

**British Library Cataloguing-in-Publication Data**
A catalogue record for this book is available from the British Library.

Published in 2002 by Imperial College Press

The designation of geographical entities in this book, and the presentation of material, do not imply the
expression of any opinions whatsoever on the part of the publisher, the editors, authors, or any other
participating organisations concerning the legal status of any country, territory, or area, or of its
authorities, or concerning the delimitations of its fronteirs or boundaries.

This book is printed on recycled paper.

Printed by FuIsland Offset Printing (S) Pte Ltd, Singapore

# Contents

# Contributors to this volume

**Penny Anderson**
Penny Anderson Associates,
Park Lea, 60 Park Road, Buxton, Derbyshire SK17 6SN, UK

**Penelope Angold**
School of Geography, University of Birmingham,
Edgebaston, Birmingham B15 2TT, UK

**Michael Ashmore**
Department of Environmental Science, University of Bradford,
Bradford, West Yorkshire BD7 1DP, UK

**Jill Barton**
Head of Conservation, Surrey Wildlife Trust,
School Lane, Purbright, Woking, Surrey GU24 0JN, UK

**Catherine J Bickmore**
Catherine Bickmore Associates,
LFG - 5, Lafone House, 11-13 Lethermarket Street, London SE1 3HN, UK

**Henri Bussink**
Alterra Landscape Ecology,
PO Box 47, 6700 AA Wageningen, THE NETHERLANDS

**John A Burton**
Old Mission Hall, Sibton Green, Saxmundham, Suffolk IP17 2 JY

**Helen Byron**
Planning Policy Officer, Royal Society for the Protection of Birds,
The Lodge, Sandy, Bedfordshire SG19 2DL, UK

**David F Cutler**
Honorary Research Fellow, Jodrell Laboratory, Royal Botanic Gardens - Kew,
Richmond, Surrey TW9 3DS, UK

**David G Dawson**
Greater London Authority,
Marsham Street, London SW1P 3PY, UK

**John M Dickson-Simpson**
Editor, Transport News Services,
Transport Press Services, 38 Portobello Road, London W11 3DH, UK

**Callum Findlay**
Environment Department, Surrey County Council,
County Hall, Kingston upon Thames, Surrey KT1 2DY, UK

**Rien F Foppen**
Alterra Landscape Ecology,
PO Box 47, 6700 AA Wageningen, THE NETHERLANDS

**Carol Hatton**
Planning Officer, WWF UK,
Panda House, Weyside Park, Godalming, Surrey GU7 1XR, UK

**Rodger M Jones**
44, Park Street Lane, Park Street, St. Albans, Hertfordshire AL2 2JB, UK

**Antony Juniper**
Director Designate, Friends of the Earth,
26-28 Underwood Street, London N1 7JQ, UK

**Derek R Langslow**
*Former Chief Executive, English Nature*
4 Engaine, Orton Longueville, Peterborough PE2 7QA, UK

**Thomas E S Langton**
Herpetofauna Consultants International Ltd.,
Triton House, Bramfield, Halesworth, Suffolk IP19 9AE, UK

**William Latimer**
WS Atkins Environment,
Television House, Mount Street, Manchester M2 5NT, UK

**Clive Livingstone**
Divisional Director, Mott MacDonald,
48-52 Andover Road, Winchester, Hampshire SO23 7BH, UK

**Alan R Outen**
*Formerly at Hertfordshire Biological Records Centre, Hertford County Hall*
15 Manor Close, Clifton, Shefford, Bedfordshire SG17 5EJ, UK

**Marian Reed**
Wessex Environmental Associates,
4 Prospect Place, Grove Lane, Redlynch, Salisbury, Wiltshire SP5 2NT, UK

**Rien Reijnen**
Senior Researcher: Landscape Ecology, Alterra Landscape Ecology,
PO Box 47, 6700 AA Wageningen, THE NETHERLANDS

**Antony Sangwine**
Head of Profession & Group Manager of Horticulture & Nature Conservation,
Horticulture & Nature Conservation Division, Highways Agency
St Christopher House, Southwark Street, London SE1 0TE, UK

**William Sheate**
Centre for Environmental Technology, TH Huxley School of Environment,
Earth Sciences and Engineering, Imperial College of Science, Technology & Medicine Prince
Consort Road, London SW7 2AZ, UK

**Hugh B Wenban-Smith**
*Former Director of National Roads Policy*
79 Rodenhurst Road, London SW4 8AF, UK

**Rowley Snazell**
Centre for Ecology & Hydrology, Natural Environment Research Council
Winfrith Technology Centre, Winfrith Newburgh, Dorchester, Dorset DT2 8ZD, UK

**Jeremy A Thomas**
Centre for Ecology & Hydrology, Natural Environment Research Council
Winfrith Technology Centre, Winfrith Newburgh, Dorchester, Dorset DT2 8ZD, UK

**Geesje Veenbaas**
Road and Hydraulic Engineering Division (DWW), Ministry of Transport and Public Works
PO Box 5044, 2600 GA Delft, THE NETHERLANDS

**Lena K Ward**
Centre for Ecology & Hydrology, Natural Environment Research Council
Winfrith Technology Centre, Winfrith Newburgh, Dorchester, Dorset DT2 8ZD, UK

**Philip Wilson**
Wessex Environmental Associates,
4 Prospect Place, Grove Lane, Redlynch, Salisbury, Wiltshire SP5 2NT, UK

**Len Wyatt**
67 Beckington Road, Knowle, Bristol BS3 5ED, UK.

**Jules Wynn**
Ecologist, WS Atkins
79 Mosley Street, Manchester M2 3LQ, UK.

**Graham Wynne**
Chief Executive, Royal Society for the Protection of Birds
The Lodge, Sandy, Bedfordshire SG19 2DL, UK

# Acknowledgements

This publication is based upon an original symposium arranged by Virginia Purchon of The Linnean Society of London, with additional support from the Royal Society for the Protection of Birds, World Wildlife Fund, and Shell UK Ltd. A contribution towards the production of this volume was provided by ACO Ltd, to whom we are most grateful.

Finally we must thank Katherine Williams and her colleagues at Imperial College Press, for all their encouragement and efforts in the production of this book.

# Preface

This volume is very timely; there are growing concerns about the impact made by roads on the environment. Many of the aspects of the complex problem of siting of new roads and lessening their negative environmental effects are addressed by specialists in their subject areas. Among the topics discussed there are papers on legal aspects, transport interests, planners' and contractors' viewpoints, plant and animal ecology and innovative solutions to some of the problems that roads inevitably impose on the natural environment.

Although the meeting at which these papers were presented was held at the Linnean Society of London in 1998, each author has been given the opportunity to make such revisions as necessary to reflect current circumstances. Consequently the information presented is up to date, and this volume represents their views at the time of writing this preface.

The chapters vary in length. Some of the posters presented at the meeting were not ephemeral, and made contributions that have a more lasting value. These are published here with the updated papers. References are provided where they are not given in full in the text.

This has been a fascinating volume to edit. Quite often symposia are of specialist interest, and not so appropriate for a wider audience. In the case of the subject matter covered here, interest is very wide, since it concerns the quality of life of us all. We hope that others will find the topics covered as informative and stimulating as we have in editing the work.

David Cutler & John Burton
*September, 2001*

# Wildlife and roads:
# A Government perspective

## H B WENBAN-SMITH

The Government's aims are twofold: first, to reaffirm their commitment to the *Biodiversity Action Plan*, published in 1994; and secondly to relate this to the Department's current work on developing an integrated transport policy. This paper will describe some of the new work on assessing road schemes, to show how environmental considerations are being given greater weight in the assessment process. It is pleasing that English Nature, who have made valuable contributions to this work, are represented in this volume.

There are still strong concerns about the effect of roads and road traffic among ecologists, despite the massive scaling-down of the road programme in recent years. Scanning the papers in this publication, there are hopeful developments also: environmental impact assessment is now built into the consideration of all major transport developments; and where intrusion into important sites is unavoidable, mitigation and compensation are actively pursued: the example of successful habitat creation for butterflies beside the M3 at Twyford Down is among the good news stories.

The Government's high level objectives are a strong economy, a sustainable environment and an inclusive society. The Government's Election Manifesto stressed the importance of policies, which combine environmental, economic and social objectives. Achieving all these objectives is what sustainable development is about. One of the Government's key initiatives to put this commitment into practice is the development of an integrated transport policy.

In June 1997, the Deputy Prime Minister (John Prescott) announced a fundamental review of transport policy. This reflected, *inter alia*, the Government's conviction that the consequences of the predicted growth of traffic would be unsustainable. The task is to find a better way forward.

## THE BIODIVERSITY ACTION PLAN

In carrying out this work, the new Government has been very mindful of the previous Government's commitment to biodiversity. As you will know, in 1992 the UK was one of more than 150 countries to sign the Convention on Biological Diversity following the Rio Earth Summit. Article 6 of the Convention commits signatories to *"...develop national strategies, plans or programmes for the conservation and sustainable use of biological diversity"*.

The UK choose to meet this commitment through the preparation of the *Biodiversity Action Plan* which was published in 1994. The UK Biodiversity Steering Group, chaired by the then DOE, but including a wide range interests from inside and outside Government, took this work forward and published their report in December 1995. The previous Government's response to the Steering Group report in May 1996 endorsed its aims and main proposals.

The Plan emphasised the significance of the growth in demand for road transport for biodiversity and highlighted the impact of road construction on wildlife habitats. It committed the Government to *"... take full account of the implications of biodiversity..."* when formulating and reviewing transport policies; and to *"...ensure that biodiversity objectives are fully incorporated into assessment and benefits of new transport infrastructure..."* The importance of the land use planning system was also emphasised. The new Government is no less committed to these objectives.

## OBJECTIVES OF INTEGRATED TRANSPORT STRATEGY

The Government want to see:

- A safe and efficient transport system which will maintain and enhance UK competitiveness.
- A better, more integrated public transport system together with better arrangements for walkers and cyclists.
- Better and more strategic integration of transport and land use planning.
- A more environmentally sustainable transport system.

It is under the last of these headings that biodiversity considerations particularly come in.

## HOW THE ROADS REVIEW FITS IN

The Roads Review is an integral part of the development of the Government's Integrated Transport Policy. It is about the role which trunk roads should play alongside other modes in an integrated and sustainable transport policy. It is not about simply hacking away at what is left of the inherited road programme. We need to take a much broader view, looking at all modes, making best use of the infrastructure we already have and targeting improvements in the light of the new Government's objectives. It is an ambitious task. What is very encouraging is the degree to which there is a public consensus in favour of a new approach.

Unlike previous reviews, this one is not being conducted behind closed doors in Whitehall. The Government has deliberately set out to seek a wide range of views: local authorities, environmentalists, transport operators, representative bodies and the general public. Many people gave their views, in response to the Department's Consultation Document *What Role for Trunk Roads in England?*.

During the Roads Review, the Department has been looking at ways of reducing the environmental impact of trunk roads. We have been considering what weight should be given to the environment in trunk road policy. How best to strike the balance between economic benefits and environmental protection, and between improving the human environment on the one hand and sustaining the natural environment on the other.

The aim is to ensure that decision making gives proper weight to the environment and is less narrowly focused. This is being done through the development of new approaches to appraisal (see below).

It would not be fair to say that environmental considerations were previously ignored. Such issues have always featured large at public inquiries and in Inspector's reports. But it is only recently that an Environmental Impact Assessment has been a legal requirement and past decisions were often perceived to have placed too much weight on economic benefits as quantified in the COBA analysis. The wider policy implications of individual decisions did not always seem to get the same attention.

## NEW APPRAISAL FRAMEWORK

### The five criteria

To address this concern, officials have been seeking to develop a New Appraisal Framework, starting from the five criteria laid down by Ministers in the Consultation Document:

- *INTEGRATION*: working together across modes and with land use planning.
- *ACCESSIBILITY*: enabling all to get to where they need to go more easily.

- *SAFETY*: reduce deaths and injuries.
- *ECONOMY*: support jobs and prosperity.
- *ENVIRONMENT*: both natural and built.

## New Appraisal Framework sub-criteria

These criteria have been broken down into various sub-criteria.

- *INTEGRATION*: Public transport - local plans - regional plans.
- *ACCESSIBILITY*: Public transport - severance - pedestrians and cyclists.
- *SAFETY*.
- *ECONOMY*: Journey times - VOCs - reliability - regeneration.
- *ENVIRONMENTAL IMPACT*: Noise - air quality - landscape - ecology - cultural heritage - water quality.

Since this paper was first written the new appraisal framework has been further developed and refined (see DTRE, 1998a,b; DTRE, 2000).

The process of assessment has started by taking existing schemes as evidence of the existence of a transport problem. But there has been no presumption that a road scheme is the right answer. Other options for dealing with the problem are being examined. If a road scheme emerges as still the most promising way forward, it is then appraised again against the five criteria. What the new approach enables us to do is to look across the various schemes in a consistent way. It summarises the key information about each scheme, including both monetised and non-monetised costs and benefits, with both qualitative and quantitative information. The aim is to ensure that all factors are given due weight in the eventual decision.

The statutory advisers, the Countryside Commission, English Heritage, English Nature and the Environment Agency are all contributing to this work. We have invited them to say how we should view the factors for which they have particular responsibility, e.g. landscape impact is for the Countryside Commission to comment on; ecology is for English Nature. We also take account of the views expressed in the consultation process.

## Biodiversity

Biodiversity comes into the appraisal process in various ways:

- *Strategic planning:* Biodiversity gets early consideration in the EIA. Those with experience in this field will know how much effort already goes into this. But we can all work together to make it even better.

- *Consultation with relevant bodies:* Bringing together the Environment and transport departments in DETR has helped. We are bringing the statutory advisors (EN, CC and EH) much more fully into the appraisal process now and this helps to ensure that their concerns are given proper weight in the decisions that are reached.

- *Special Consideration for designated sites*: The more important ones now benefit from the protection of the relevant European Directives. But in addition, Ministers have already indicated that they wish there to be a strong presumption against building roads through designated environmentally sensitive sites. This is one of the issues being closely examined in the current review.

- *Mitigation*: Where intrusion cannot be avoided, mitigation and/or compensation can make a big difference to the acceptability of a proposed scheme. Measures can include new habitat creation, species relocation, habitat management commitments, protection of similar habitats at other locations and so on. It seems that there is real scope in this situation to look for 'win/win' solutions, which both solve the transport problem and achieve real environmental benefits.

- *Scheme design*: The Highways Agency Design Manual for Roads and Bridges covers design issues very comprehensively as well as mitigation. DMRB will be updated and revised in the wake of the Roads Review and this will be covered by Tony Sangwine in this volume.

- *Continuing research and monitoring*: This provides the data and analysis which feeds back into the appraisal process described.

## CONCLUSION

This paper has indicated the ways in which environmental considerations are being given much greater weight in the current transport reviews. No doubt there is much more that can be done and I hope that among the outcomes of consultation will be:

- Authoritative advice on the problems that merit priority attention.

- Improvements in the methods we use to assess such problems; and better advise on the measures we can take to minimise environmental impact or even enhance the environment.

# REFERENCES

**DETR. 1998a.** *Understanding the New Approach to Appraisal.* London: Department of the Environment, Trade and the Regions.

**DETR. 1998b.** *Guidance on the New Approach to Appraisal.* London: Department of the Environment, Trade and the Regions.

**DETR. 2000.** *Guidance on the Methodology for Multi-modal Studies.* London: Department of the Environment, Trade and the Regions.

# 2

# Requirements of EU and UK
# wildlife legislation

## C HATTON

*Let's say goodbye to hedges*
*And roads with grassy edges*
*And winding country lanes;*
*Let all things travel faster*
*Where motor-car is master*
*Till only speed remains*

From *Inexpensive Progress* John Betjeman

There are two European Union Directives primarily concerned with the conservation of biodiversity – the *Birds Directive* and the *Habitats Directive*. In the UK the primary statute concerned with the conservation of biodiversity is the *Wildlife and Countryside Act 1981*, although other Acts (such as the Town and Country Planning Acts) and agreements (such as the *Biodiversity Action Plan*) have a significant impact on biodiversity.

This chapter is concerned with the implications of the *Birds Directive*, the *Habitats Directive* and the *Wildlife and Countryside Act 1981* for roads and road planning. It sets out the broad requirements of each statute in terms of site designation and protection, species protection and the wider countryside. Where appropriate, linkages are made with non-legislative measures such as the *Biodiversity Action Plan* and the sustainable development strategy. The paper concludes with recommendations for road policy, planning and practice.

Historically the needs of biodiversity have not been central to, or even integrated into, transport planning. The forthcoming Transport White Paper is an appropriate place for the Government to reconcile the needs of biodiversity within an integrated transport system. WWF-UK is a member of the Transport and Biodiversity Working Group (TBG) which also includes the Royal Society for the Protection of Birds (RSPB), The Wildlife Trusts and English Nature. The Group has devised a series of biodiversity objectives for the transport sector and those relating to EU and UK wildlife legislation are included in this paper.

## EU WILDLIFE LEGISLATION

### The Birds Directive

EC Council Directive on the Conservation of Wild Birds (79/409/EEC): The *Birds Directive* was agreed on 2 April 1979. The general objective of the Directive is *"...the conservation of all species of naturally occurring birds in the wild state in the European territory of the Member States to which the Treaty applies"* (Article 1(In lit. 1998)). It covers the protection, management and control of these species applying to birds, their eggs, nests and habitats.

Articles 2 and 3 require Member States to take measures to preserve a sufficient diversity of habitats for all species of wild birds naturally occurring within their territories in order to maintain populations at ecologically and scientifically sound levels. Whilst allowing Member States to take account of economic and recreational requirements, the Directive therefore seeks to provide a general level of protection for all wild bird species within the territory of the European Community.

Article 4 requires Member States to take special measures to conserve the habitat of certain particularly rare species and of migratory species listed in Annex I. There are 175 species listed in Annex I and those occurring in the UK include the whooper swan *Cygnus cygnus*, osprey *Pandion haliaetus*, corncrake *Crex crex*, stone curlew *Burhinus oedicnemus*, and Dartford warbler *Sylvia undata*. Member States are required to classify Special Protection Areas (SPA) for Annex I species, taking into account their protection requirements in the geographical sea and land area where this Directive applies. Member States are required to take similar measures for regularly occurring migratory species not listed on Annex I as regards their breeding, moulting and wintering areas and staging posts along their migration routes.

Until 1995, Special Protection Areas were protected by Article 4 of the *Birds Directive* which required Member States to take appropriate steps to avoid pollution or deterioration of habitats or any disturbances affecting the birds (Article 4 (Case-44, 1995)). Since 1995, the protection afforded to SPAs under Article 4 (Case-44, 1995) has been replaced by the provisions of Article 6 of the *Habitats Directive*.

The UK has classified 169 SPAs covering over 708,890 hectares (In lit. 1998). No sites is classified as an SPA unless it is first notified as a Site of Special Scientific Interest (SSSI) under Section 28 of the *Wildlife and Countryside Act 1981* which was passed to transpose the *Birds Directive* into UK law.

The UK is also signatory to the Convention on Wetlands of International Importance especially as Waterfowl Habitat (the Ramsar Convention – named after the town where the first conference was held in 1971). The broad objectives of the Convention are to stem the progressive encroachment on, and loss of, wetlands and to promote their wise use. The UK has listed 120 Ramsar sites covering a total of 491,646 hectares (In lit., 1998). Whilst not protected by European law, as a matter of policy Ramsar sites are given protection equivalent to SPAs.

## The Habitats Directive

EC Council Directive on the Conservation of Natural Habitats and of Wild Fauna and Flora (92/43/EEC): The *Habitats Directive* was agreed on 21 May 1992, more than a decade after legislation for the conservation of wild birds. The general objective of the Directive is *"...to contribute towards ensuring biodiversity through the conservation of natural habitats and of wild flora and fauna in the European territory of the Member States to which the Treaty applies"* (Article 2 (In lit., 1998)).

The Directive is divided into two main sections – Articles 3-9 covering the conservation of natural habitats and habitats of species, and Articles 12-16 covering the protection of species outside protected areas. Other Articles cover the wider countryside, monitoring, information, research and supplementary provisions. The central tenet of the Directive is set down in Article 3 (In lit., 1998) which requires Member States to establish a Community wide network of Special Areas of Conservation (SAC) which would maintain at (or restore to) 'favourable conservation status' the habitats and species of Community importance listed in Annexes I and II.

There are 168 habitat types listed on Annex I and 193 animal and 432 plant species listed on Annex II. Annex I specifies the types of habitats to be protected, of which 75 occur in Great Britain including raised bogs, old oak woodlands with holly and hard fern, Caledonian pine forest, estuaries, caves and lowland dry heathlands. Annex II identifies animal and plant species which, due to their special interest, require the designation of SACs for their protection. There are 40 species occurring in GB, including the otter *Lutra lutra*, harbour porpoise *Phocoena phocoena*, great crested newt *Triturus cristatus*, stag beetle *Lucanus cervus*, fen orchid *Dactylorhiza incarnata* and

shore dock *Rumex rupestris*. Certain habitats and species listed on Annexes I
and II are identified as a priority (indicated by an asterisk (\*)) for protection
and funding under certain circumstances. Examples of priority habitats in
the UK include active raised bogs, Caledonian forest and lagoons. There is
only one priority species in the UK, the western rustwort *Marsupella
profunda*. Detailed scientific criteria for site selection are listed in Annex III.
Annex IV lists species in need of strict protection and Annex V lists species
which require management measures for their exploitation.

'Favourable Conservation Status' (FCS) is defined as *"...when the
species population and range is stable (or increasing) and there is a
sufficiently large area of habitat available to maintain its population on a
long-term basis"* (Article 1(j)). The definition is taken from the Bonn
Convention of 1979 (on the conservation of migratory species), in which it
appeared for the first and apparently the only time prior to the adoption of
the Directive (2). Interpreting 'favourable conservation status' is proving
problematic, however, it has been suggested that it will have to be
considerably higher than the minimum viable population (De Klemm, 1995).

The *Habitats Directive* is implemented in the UK through the *The
Conservation (Natural Habitats, &c.) Regulations 1994*. The Regulations
were supplemented in England by the publication of Planning Policy
Guidance: Nature Conservation (PPG 9). In March 1995, the Government
consulted the public on 280 Candidate Special Area for Conservation
(cSACs) and a further 42 (plus the criteria for their selection) followed on 1
October 1997. The Government has sent 262 cSACs covering 1,526,817
hectares (about 3% of the UK territory) (Anon, 1998a) to the European
Commission. As a matter of Government policy, candidate SACs are now
protected under Article 6 of the *Habitats Directive*.

## Natura 2000

Together, Special Areas of Conservation and Special Protection Areas
make up a pan-European network of protected areas termed *Natura 2000*.

### Natura 2000 – Designation

Recent judgements of the European Court of Justice (Anon, 1996) and
the House of Lords (Anon, 1997a) have confirmed that SPA boundaries
must reflect only ecological factors. Thus, it would appear that the provision
in Article 2 of the *Birds Directive* to take account of economic and
recreational requirements into account does not extend to site identification.

It remains to be seen whether this principle applies equally to SACs.
Whilst the *Habitats Directive* also allows Member States to take account of

*"... economic, social and cultural requirements and regional and local characteristics"* (Article 2(3)) it seems reasonable for the designation of SACs to be consistent with SPAs. It is hoped that the current WWF-UK/RSPB case against the Scottish Office, Scottish Natural Heritage and The Highland Council concerning a proposed funicular railway development in the Cairngorms will clarify this point.

In the interim, it would appear that the boundaries of a number of cSACs reflect economic considerations. For example, the boundary of the Orton Brick Pits in Peterborough, a cSAC for the great crested newt *Triturus cristatus*, clearly follows the route of the proposed A15 which will serve Hampton New Town. By excluding qualifying areas from the cSAC, statutory procedures can be avoided including those concerning assessment (Article 6), the evaluation as to whether the project must be carried out for imperative reasons of overriding public interest (Article 6(4)) and compensatory measures.

### *Natura 2000 – Protection*

Measures to be taken in respect of *Natura 2000* sites are set out in Article 6 (1-4) of the *Habitats Directive* which has replaced Article 4(4) of the *Birds Directive*. These measures include management plans where appropriate (Article 6(1)) and an environmental assessment of all non-management projects which may have a significant effect on *Natura 2000* sites (Article 6(3)). It is WWFs opinion that such an assessment will not be satisfied by an Environmental Impact Assessment (EIA). An assessment under Article 6(3) must evaluate whether the plan or project will affect the favourable conservation status of the habitat or species in question at the regional, national and European levels and will, therefore, be more specific than an EIA. Furthermore, the opinion of the general public may be required.

If the assessment shows that there will be a negative ecological impact s a result of the project, it may only proceed *"...for imperative reasons of overriding public interest, including those of a social or economic nature"* but only where no alternative exists and, where appropriate, an opinion has been sought from the general public. In this eventuality, compensatory measures must be taken to ensure that the overall coherence of *Natura 2000* is protected (Article 6(4)). WWF urges caution with regard to the translocation of species and habitats. A recent WWF-UK report *A Moving Story – Species and community translocation in the UK: a review of policy, principle, planning and practice* (Gault, 1997) concludes that translocation is a relatively unknown science and should only be regarded as a form of mitigation where an irreplaceable site faces destruction. It should never be viewed as available option at the planning stage.

Where the site in question supports a habitat and/or species listed as a priority (*) on the *Habitats Directive*, development proposals can only proceed for considerations relating to *"...human health or public safety, to beneficial consequences of primary importance for the environment or, further to an opinion from the Commission, to other imperative reasons of overriding public interest"*. In essence this provides a two-tier system; proposals affecting SACs supporting non-priority species or habitats may in certain circumstances be permitted on social or economic grounds. Proposals affecting SACs supporting priority species and habitats may only be permitted on grounds of public health or safety or imperative reasons of overriding public interest. For the purpose of considering development proposals affecting them, potential SPAs and candidate SACs should be treated in the same way as classified SPAs and designated SACs (Anon, 1994). In this respect, the passage of the *Habitats Directive* has weakened existing case law on the *Birds Directive*. In 1991, the German 'Leybrucht Dykes' case (Anon, 1991) ruled that development affecting SPAs must not be sanctioned unless required for reasons of human health or safety. Although the protection afforded to SPAs under Article 6 of the *Habitats Directive* is not clear, the European Commission's view (Cashman L., Legal Unit, DG XI, European Commission, *pers. comm.* 1998) is that proposals which may affect them are subject to the less stringent test in Article 6 (i.e. imperative reasons of overriding public interest, including those of a social or economic nature). One of the Biodiversity Objectives drawn up by the TBG concerns the protection of internationally important sites, including SPAs, SACs and Ramsar sites. The TBG recommend that there should be no damage to, or destruction of, internationally important sites unless proposals are necessary for reasons of human health or safety or for other imperative reasons of overriding public interest.

Finally, it is also worth noting that the level of protection to be afforded to Annex I habitats outside SACs is far from clear. Guidance, perhaps in the form of a revised PPG 9, is necessary.

## The protection of species under EU law

Articles 12 to 16 of the *Habitats Directive* concern the protection of species outside SACs. Article 12(1) requires Member States to establish a system of strict protection for animal and plant species listed in Annexes IVa and IVb, prohibiting deliberate capture, killing and disturbance of these species in the wild and the destruction of their breeding or resting places.

Member States are also required to establish a system to monitor the incidental capture and killing of animal species listed in Annex IVa and, in the light of information gathered, undertake further research or conservation measures to ensure that there is not a significant negative impact on the species concerned (Article 12(4)).

Article 16 enables Member States to derogations from the requirements of Articles 12, 13, 14 and 15 (a) and (b) in certain circumstances. Article 16* allows for a derogation in the interests of public health and safety, or for other imperative reasons of overriding public interest, including those of a social or economic nature *"... where there is no satisfactory alternative and the derogation is not detrimental to the maintenance of the species' favourable conservation status"*. Thus, species listed on Annexes II and IVa outside SACs should not normally be *"... captured, killed or deliberately disturbed, particularly during breeding, rearing, hibernation and migration"*. Furthermore, Member States are required to prohibit the destruction of their breeding sites or resting places.

In this respect, *The Conservation (Natural Habitats &c.) Regulations 1994* are inconsistent with the requirements of the *Habitats Directive*. Regulations 40(3)(c) and 43(4) provide a loophole to the requirements of Article 12 for animals and plants respectively where it can be shown that the act was *"...an incidental result of a lawful operation and could not reasonably have been avoided"*. This wording, taken directly from Section 10(c) of the *Wildlife and Countryside Act 1981*, has no origin in the *Habitats Directive* and should be removed. The consequence of the UK government's interpretation is that it enables the habitat of an Annex IVa species to be destroyed if it is the incidental result of a lawful operation without satisfying the more stringent tests set out in Article 16 of the Directive. This includes the absence of a satisfactory alternative and the need for an assessment as to whether the proposal is detrimental to the maintenance of the population at FCS. A practical example of this is the Newbury Bypass in Berkshire which destroyed important habitat for the dormouse, a species listed on the *Habitats Directive*.

The TBG is concerned about the UK's interpretation of the *Habitats Directive* and has drafted an objective, which we believe more accurately reflects its intention. The Group recommend that there should be a strong presumption against damage to, or destruction of, habitats listed in Annex I and species listed on Annex II and IVa (including the habitat of said species outside SACs) of the *Habitats Directive*. Consent should only be granted where it can be demonstrated by the applicant that the proposal will not prejudice the achievement of 'favourable conservation status' (following the

definition given in the *Habitats Directive*) at the regional, national and
Community level.

## The wider countryside and EU law

Protected areas cannot exist in isolation and will not, in themselves,
deliver the conservation of biodiversity in accordance with international
commitments and obligations. Other significant areas include land of
importance as buffers or corridors which are essential for the recovery of
species and programmes of habitat restoration. Both the *Birds* and *Habitats
Directives* recognise the need for action in the 'wider countryside', however,
both also fall short of requiring measures to conserve and enhance features
outside *Natura 2000*.

Article 4(4) of the *Birds Directive* encourages Member States to
"*...strive to avoid pollution or deterioration of habitats*" outside SPAs.
Similarly, Article 10 of the *Habitats Directive* urges Member States to
modify land use and planning policies to improve the ecological coherence
of *Natura 2000* and encourages the management of landscape features which
are of major importance for wild fauna and flora. Such features include
rivers and their banks, hedgerows or other field boundary systems and small
ponds or woods. Accordingly, WWF-UK urges local planning authorities
and relevant bodies to identify land of importance as buffer and/or corridor
on appropriate plans and to impose a presumption against damaging
development affecting such areas. WWF-UK also urges the Department of
the Environment, Transport and the Regions to apply Strategic
Environmental Assessment to any future integrated transport system. This
would enable DETR to evaluate the implications of regional and national
transport policies, not only on *Natura 2000*, but also on features of
importance in the wider countryside covered by Article 10.

## UK LEGISLATION

The *Wildlife and Countryside Act 1981* relies heavily on the 'voluntary
principle' (i.e. the goodwill of landowners not to damage important sites). It
is not a powerful statute and its inadequacies have been detailed in a number
of governmental and non-governmental publications (Rodwell, 1991). The
Labour government is committed to introducing new wildlife legislation and
WWF was pressing for the inclusion of a Bill in the 1998 Queen's speech.

Land of special interest for nature conservation may be notified as a Site
of Special Scientific Interest (SSSI) under Section 28 of the *Wildlife and
Countryside Act 1981* (or as Areas of Special Scientific Interest (ASSI) in

Northern Ireland). WWF-UK contends that all SSSIs are of national importance, each making a significant contribution to the national series of sites. There were just over 4,000 SSSIs in England covering 955,000 hectares on 19 September 1997 (Anon, 1997b).

The TBG believe that the principles controlling development in the countryside should apply more strictly in SSSIs. Such proposals should not be approved, except in exceptional circumstances i.e. where it can be shown that the proposal is in the overriding national interest and could not be achieved in any other way. It is for the applicant to demonstrate why permission should be granted and, through the assessment of alternatives, to demonstrate why such alternatives are not appropriate.

The 1981 Act also protects species outside SSSIs. Broadly speaking it is an offence to intentionally kill, injure, take or sell wild animals or intentionally disturb them or damage their places of shelter. All wild birds, their nests and eggs are protected by law and there are special penalties for harming certain species. Finally, specially protected plants (listed in Schedule 8 of the Act) must not be picked or sold. Uprooting any wild plant is illegal. The insertion of Section 10(c) in the 1981 Act legalises many of the above operations if they are the incidental result of a lawful operation and could not reasonably have been avoided. This Section therefore legalises the destruction of SSSIs and protected species by roads where planning permission constitutes a lawful operation.

The *Wildlife and Countryside Act 1981* is not the only statute concerned with the protection of wildlife in the UK. The *Protection of Badgers Act 1992* prohibits the killing, injuring or taking of a badger and damage to a badger sett. The Act is similarly subject to the 'incidental result of a lawful operation' proviso which legalises the destruction of badger setts by activities given planning permission, such as road schemes. For reasons of brevity, other statutes are not examined here.

The *Biodiversity Action Plan* (Anon, 1994/95) is an essential contribution to the UK's implementation of the Biodiversity Convention. Transport, and specifically roads, threaten 11 of the 16 species and 8 of the 38 key habitats identified as priorities for action in *'Biodiversity: The UK Steering Group Report'* (Anon, 1995). The UK Steering Group called for 'the implications of the Biodiversity Convention [to be] considered by individual Departments'.

The TBG urge the government to address this issue in the forthcoming Transport White Paper. With respect to individual projects, the Group recommends that there should be a presumption against the destruction of, or damage to, habitats, species and habitats of species listed in the *Biodiversity Action Plan*.

## BIODIVERSITY AND SUSTAINABILITY

The opening preamble of the *Habitats Directive* sets out in general terms he aims and objectives of the Directive. Most specifically, in addition to seeking to promote the maintenance of biodiversity, the Directive strives to make a contribution to the general objective of sustainable development. However, there is nothing more specific about this in the Articles of the *Habitats Directive*.

A number of underlying principles for sustainable use are set out in *Biodiversity Challenge – an agenda for conservation in the UK* (Anon, 1994/95) and those of most relevance to the transport sector include:

- Conservation of biodiversity should be an integral part of all government programmes, policy and actions.

- The precautionary principle should guide all decisions that might cause environmental damage.

- Environmental appraisal, as well as financial appraisal, should be widely undertaken at both strategic and project level.

- Critical natural capital must be conserved.

WWF-UK urges the government to address these recommendations.

## CONCLUSION

The EU *Birds* and *Habitats Directives* provide a higher level of protection for a relatively small number of SSSIs, some of which (SACs supporting priority interests) can only be damaged or destroyed for issues relating to human health or safety or other imperative reasons of overriding public interest. This is the strongest protection afforded to any UK designated area. There is, however, continued concern that effort and resources will be focused on those SSSIs protected by EU law, with worrying consequences for the majority of the SSSI series. This must not be allowed to occur.

The Directives' attempts to stem neglect and deterioration would be a significant step forward if recognised and implemented in the UK. Whether the Directives' good intent is realised will depend upon whether the UK government is prepared to move away from placing an unduly narrow interpretation on the Articles of the Directive.

It is difficult to predict whether habitats listed on Annex I will have any added protection outside SACs. Species listed on Annexes II and IVa will, however, receive some added protection, although the extent of this remains unclear. The Government must ensure that the provisions of Article 12 (species protection) are given due weight by decision-makers – both at the strategic and project level. Similarly, the protection of wider countryside features will depend upon clear advice, reinforcing the importance of Planning Policy Guidance.

The effectiveness of EU law also depends upon the proper transposition and wide interpretation of the Directive. If in doubt about interpretation, decision-makers should refer to the intention of the Directives as set out in the Preamble. Finally, it is important to ensure that EU law is properly enforced – legislation may be robust but it must be applied if it is to be of use.

The *Wildlife and Countryside Act 1981* remains a blunt instrument and it is largely public policy, not legislation, which has reduced the impact of roads on protected sites. New wildlife legislation to replace the 1981 Act is urgently needed.

## RECOMMENDATIONS

WWF-UK urges the government, transport planners and practitioners.

### For EU legislation

- *Natura 2000 Designation* – ensure that the boundaries of all *Natura 2000* sites reflect ecological factors alone.

- *Natura 2000 Protection* – recognise that an 'appropriate assessment' of any plan or project likely to have a significant effect on a *Natura 2000* site (Article 6(3)) should extend beyond Environmental Impact Assessment to comprise a more detailed assessment of the impacts of the plan or project on favourable conservation status at relevant levels.

- *Natura 2000 Protection* – strictly apply the tests set out in Article 6 of the *Habitats Directive* regarding damage to, and destruction of, *Natura 2000* sites. The TBG recommend that there should be no damage to, or destruction of, internationally important sites unless proposals are necessary for reasons of human health or safety or for other imperative reasons of overriding public interest.

- *Species Protection* – clarify the protection given to species listed on Annexes II and IVa of the *Habitats Directive* outside SACs. The TBG recommend that there should be a strong presumption against damage to, or destruction of, habitats listed in Annex I and species listed on Annex II and IVa (including the habitat of said species outside SACs) of the *Habitats Directive*.

- *Habitat Protection* – clarify the protection afforded to Annex I habitats outside SACs.

- *Wider Countryside* – ensure the maintenance of features of interest in the wider countryside.

## UK legislation and agreements

- *Wildlife and Countryside Act 1981* – include new wildlife legislation replacing the 1981 Act in the 1998 Queen's Speech.

- *SSSIs* – Protection – ensure that principles controlling development in the countryside be applied more strictly in nationally important areas – such proposals should not be approved unless it can be shown that the proposal is in the overriding national interest and could not be achieved in any other way.

- *Biodiversity Action Plan* – introduce a presumption against the destruction of, or damage to, habitats, species and habitats of species listed in the government's *Biodiversity Action Plan*.

- *Biodiversity and sustainability* – ensure that: the conservation of biodiversity is an integral part of all government programmes, policy and actions; the precautionary principle guides all decisions that might cause environmental damage; environmental appraisal, as well as financial appraisal, is widely undertaken at both strategic and project level; and critical natural capital is conserved.

## REFERENCES

**Anon. 1991.** Case C-57/89 Commission of the European Communities v Federal Republic of Germany, judgement delivered on 28 February 1991.

**Anon. 1994.** *Planning Policy Guidance: Nature Conservation (PPG 9).* Department of the Environment October 1994.

**Anon. 1995.** Biodiversity: The UK Steering Group Report HMSO 1995.

**Anon. 1996.** Case-44/95 R v Secretary of State for the Environment ex parte RSPB, judgement delivered on 11/7/96).

**Anon. 1997a.** R v Secretary of State for the Environment ex parte RSPB Judgement Order of the House of Lords delivered on 13/3/1997.

**Anon. 1997c.** English Nature Press Release *Health Check on SSSIs as the 4,000th SSSI is notified.* 19 September 1997.

**Anon. 1998a.** Eagle A. 1998 Answer to written Parliamentary Question number 603 given by Angela Eagle MP in response to a question from Cynog Dafis MP on the European Union Habitats Directive on 17 February 1998.

**Anon. 1998b.** Figures obtained from DETR, March 3, 1998.

**Butterfly Conservation, Friends of the Earth, Plantlife, Royal Society for Nature Conservation, the Wildlife Trusts' Partnership, Royal Society for the Protection of Birds and World Wide Fund For Nature UK 1994/5.** *Biodiversity Challenge I and II.* Sandy, Beds: RSPB.

**Gault C. 1997b.** *A Moving Story - Species and community translocation in the UK: a review of policy, principal, planning and practice.* WWF-UK.

**De Klemm C. (1995)** *Legal Instruments for the Protection of Wild Flora.* Plenary Paper for the Planta Europa Conference (2-8/9/95). Paris: France.

**Rowell TA. 1991.** *SSSIs - A Health Check A Review of the Statutory Protection afforded to Sites of Special Scientific Interest in Great Britain.* A Report for Wildlife Link.

# 3

# Highways and wildlife in the UK:
# the Highways Agency view

A SANGWINE

The Highways Agency is an agency of the Department of the Environment, Transport and the Regions, and was created on the first of April 1994. It has responsibility for the operation and improvement of the trunk road network in England. The network extends to nearly 6600 miles (10,000 kilometres) and includes the English motorways. It encompasses an estate which comprises chiefly of the land from edge of carriageway to the highway boundary.

The trunk road network was established by the Trunk Roads Act of 1936 and for the second time in Britain's history put a strategic road transport network under direct government control. The Romans had built such a network of roads between AD40 and 80 of similar extent (about 6500 miles) which were functional military and trade arteries. They also acted as a conduit for a group of plants, essentially the core of the many species we now regard as roadside or verge colonisers. These plants include corn cockle *Agrostemma githago*, scarlet pimpernel *Anagallis arvensis*, greater celandine *Chelidonium majus*, cotton thistle *Onopordum acanthium*, white mustard *Sinapis alba* and field wound wort *Stachys arvensis*. After the Romans left these shores no comparable road building or management was undertaken until the advent of the Turnpike Trusts in 1663 and thereafter. They charged tolls in exchange for providing a maintained highway. Sadly, they frequently failed in their duty and in 1832 Parliament gave this responsibility to the new parish authorities and a steady improvement ensued allied to the technologies of Telford and Macadam.

The trunk road network incorporated Roman roads, as in the Midlands where the A5 follows Watling Street for much of its length, as well as Turnpike roads and the drove roads that catered for the great movement of cattle and sheep between farms, fairs and markets. This element of our modern network has given us a group of wild flowers whose names betray their habitat in the 16th. and 17th. centuries. White clover *Trifolium repens*, known as sheep's gowan in some parts and its companion grass sheep's fescue *Fescua ovina* were the animals food, whilst sheep's-bit *Jasione montana* and sheep's bells or harebells *Campanula rotundifolia*, flourished in the tight sward that resulted from their grazing.

These early thoroughfares though sometimes heavily trafficked were not serious obstacles to the movement of wildlife. The 19th century railways had more impact on wildlife through the imposition of physical barriers such as lengthy embankments, deep cuttings and built infrastructure (stations, goods yards and works). Half this once intricate network has been abandoned but the remaining network can be a major barrier to wildlife especially where 750 volts direct current third rail systems are used. In south-east England as many as 2000 badgers *Meles meles* may be killed every year negotiating this hazard.

The rise of the modern highway was of course associated with the growth of motor traffic and ambitious plans were laid down in the 1930s for a network of dual carriageway roads but this was all cut short by the outbreak of the Second World War. Post was economic conditions militated against an early recommencement of such a programme. In 1956 the motorway building era began with what was to become the first part of the M6 – the Preston Bypass. Nowhere in the literature of this period can one find reference to the environmental impact of motorway construction and operation. Certainly the landscape treatment of the early motorways was confined to scattered tree planting using light standard stock and hawthorn, *Crataegus monogyna*, hedges.

This approach owes something to the advice of the Roads Beautifying Association (RBA) which guided highway planting policy with a series of recommendations for species choice and sitting. They advised planting densities of 60 trees per mile and the use of exotic shrubs such as *Cotoneaster, Berberis, Lonicera, Pyracantha* and *Veronica*. Native trees were used but so also were many exotics especially varieties of *Prunus, Acer* and *Sorbus*. The RBA disbanded in 1956.

The Landscape Advisory Committee did, to some extent replace the RBA and had a considerable influence on the quality of the landscape design of modern highways in England and Wales. The development of a better informed planting policy was founded on a programme of research carried out for the Department of Environment and Transport between 1970 and. 1985. The outcome of his was the establishment of the present techniques for planting highways using small bare root stock, usually of 450 to 600 millimetres size, pit planted and maintained in the establishment phase by spot herbicide treatment. This provides freedom from competition which is critical to sustained growth in the first three to five years. Since 1958 some 46 million trees and shrubs have been planted alongside England's trunk roads. These plantings comprise predominately native broadleaves in their natural associations. Some 14,500 hectares have been planted or incorporated (in the case of natural regeneration) in the post-war programme.

The greater part of the trunk road estate is grassland, about 15,500 hectares. This offers some of the greatest potential for wildlife since it is an undisturbed linear reserve where no pesticide is used. It forms an ecotone where habitats succeed each other and merge. The mown verge leads to the unmown grassland which borders the shrub edge of the linear woodland or mixed scrub. There will be many opportunities or niches for invertebrates, small mammals and reptiles. The diversification of the grass sward to provide greater botanical interest has been progressing slowly in recent years. Experiments with wild flower sowing and planting in the late 1980s proved the techniques for new sites. The treatment of existing swards is more difficult and is best focused on areas which are already being colonised by desirable wild flower species or offer the right combination of poor fertility and some freedom from competition. A reservoir of suitable species can be placed in an existing sward by using plants, preferably pot grown or for mass plantings plugs gown in cell systems. Cutting and removing arising at the right time of the year is a necessary maintenance operation to create the right conditions for the spread of desirable species. The Agency has a widely acclaimed Wildflower Handbook in the *Design Manual for Roads and Bridges*, which describes how to design, implement and manage wild flower swards.

Scrub and woodland take much longer to develop significant wildlife interest. The woodland edge is the first element to attract wildlife and butterflies and their larvae are found within the first ten years of planting. The use of native shrubs such as privet *Ligustrum vulgare*, dogwood *Cornus sanguinea*, blackthorn *Prunus spinosa*, sallow *Salix caprea* and holly *Ilex aquifolium*, encourage a range of butterflies. Small mammals such as the bank vole *Clethrionomys glareolus*, the short-tailed vole *Microtus agrestis* and the long-tailed field mouse *Apodemus sylvaticus,* and less frequently encountered, the harvest mouse *Micromys minutus*, are rapid colonisers of the highway estate. They attract kestrels *Falco tinnunculus*, which are well adapted to roadside hunting as they dive vertically upon their prey. The barn owl *Tyto alba*, by contrast glides across field or carriageway to hunt and is often a casualty of fast moving traffic.

An added dimension to the grassland and woodland of the roadside is the extensive wetland that serves both to drain the carriageway and safeguard receiving waters from pollutants. Open ditches and balancing pounds are found all over the network and within motorway interchanges substantial bodies of standing water attract their own wildlife.

A location that encapsulates the full range of wildlife interest is junction 16 on the M25 motorway. There are 90 acres (36 hectares) of land within the interchange between the M25 and M40 motorways. There is a mature mixed coniferous and deciduous woodland planted more than 45 years ago and incorporated in the site, two balancing ponds that hold water all the year round, ditches, scrub, young woodland (10 years old) and a variety of grassland types. The area is home to roe deer *Capreolus capreolus*, muntjac *Muntiacus reevesi*, foxes *Vulpes vulpes*, rabbits *Oryctolagus cuniculus*, coots *Fulica atra*, (they nest on the balancing ponds and have been there for 4 years), voles, mice and many woodland birds. All enjoy this habitat despite the traffic noise, vibration and lighting at night. More than 80,000 trees and shrubs have been planted in the area to screen traffic from view of scattered dwellings and integrate the junction into the landscape.

So it can be seen that the modern highway estate has an attraction to wildlife. The motorway network is particularly valuable as a conservation resource since modern agricultural practices and urbanisation have acted to discourage our native fauna and flora. These have retreated to its undisturbed reaches. Its linear nature means it provides a linking corridor with other remnant habitats such as copses, hedgerows, field margins, waste ground and scrub . Certain species are at risk because they cross motorways notably deer, badgers and foxes. The provision of dedicated or adapted crossings for these larger mammals is now part of current highway design. Fencing is provided to ensure animals use safe crossings, This can range from short plastic curved fences for amphibians to 1.8 metre high wire fences for fallow deer *Dama dama*.

In support of the design, construction and operation of the trunk roads a design manual simply titled *Design Manual for Roads and Bridges* has been developed over a period of years to reflect the latest developments and techniques. The volumes concerned with environmental assessment and environmental design are numbers 11 and 10 respectively. Volume 10 is the *Good Roads Guide* series of advice notes and this is being enlarged to cover all aspects of nature conservation and roads. In March 1997 the definitive advice on mitigating the effects of highways on badgers was published and in the summer of 1998 the Agency will publish new advice chapters on otters *Lutra lutra*, and bats. This work is underpinned by a research programme which in 1998 will be examining a range of species to determine the impact that roads have on their habitat and further guidance chapters in the manual will be produced. This programme will be focusing on dormice *Muscardinus avellanarius*, reptiles, amphibians, nesting birds and deer.

The Agency has a role to play in the UK Biodiversity Plan and is a named participant in the delivery of specific components notable among them is protection for otters.  At the Local Biodiversity Action Plan level the Agency will be involved in implementing targeted management of habitats that form part of the highways estate, or are contiguous with it.  The more complex and fragile components of the highway estate are the relocated habitats which may require long establishment periods before they achieve any stability.  They will require specialised maintenance as appropriate to the site, species composition and seasonal factors.

Measures for the protection of wildlife such as badger tunnels and fences will require inspection and maintenance as will culverts, balancing ponds, otter tunnels and ledges, bat roosts, reptile hibernaculum and trans-located habitats.

The objectives for the future are to realise the full potential of the highways estate for nature conservation whilst operating the highway safety and economically.  Mitigating effects on wildlife is a complicated process and requires a balance to be struck between the provision of habitat and protection from the deleterious effect of traffic and engineering operations. That is the challenge in the next century and the Highways Agency is pushing out the frontiers of its knowledge in the areas of nature conservation and highway operation to provide state of the art guidance on managing this mosaic of elements that form the modern highway.

Deputy Prime Minister (John Prescott) announced a fundamental review of transport policy. This reflected, *inter alia*, the Government's conviction that the consequences of the predicted growth of traffic would be unsustainable. The task is to find a better way forward.

# 4

# Statutory agencies and roads:
# English Nature

## D LANGSLOW

In 1997, the Government conducted an urgent review on the inherited Roads Programme. It decided not to build some roads, to go ahead with others and to conduct a further review of around 100 schemes. For us, the cancellation of the Salisbury By-pass was hugely welcomed. The combined forces of the statutory and voluntary movement successfully argued that the route was damaging and unnecessary. Along with the Countryside Commission, we fought hard behind the scenes to persuade Ministers to cancel the project.

We should be under no illusion about the love affair of most people with the car. Thus, if we oppose schemes designed to help many people, we may be in a minority. In looking ahead, it is vital that we increase the amount of practical, and not just rhetorical, support on issues about roads and the environment.

## ENGLISH NATURE'S DUTIES

One of English Nature's most important duties is to act as statutory advisers to Government on nature conservation in England. Thus our primary focus is to examine the ecological impact both of the roads programme as a whole and individual schemes. Our advice needs to be backed up by sound evidence with clear value statements. It is also our responsibility to examine alternative propositions on particular roads schemes which will limit or prevent damage to the environment. Thus our actions concerned with roads range from meetings with Ministers, much work with senior officials in DETR, close contact with Highways Agency at all levels and local level work with Local Authorities and local interest groups on particular schemes.

## PROSPECTS FOR CHANGE

The new Millennium offers an enticing prospect as a marker for change. Achieving the correct policies in the coming months could mean a more environmentally sustainable transport policy in the future. Let us also be clear that an effective road transport system remains essential to our established lifestyles, conveys much freight efficiently and supports the individual's freedom to travel.

Arguing for particular restrictions on groups of individuals is easy, but we all need to be wary about adopting an elitist and restrictive stance which cannot be justified widely. Many of us either owns a car or uses one regularly and enjoys interacting with wildlife. How many of us use a car as a necessary part of achieving wildlife experiences?

## THE LAST NINE YEARS

In looking at English Nature's policies towards roads, I want to begin in 1989. Frankly, it is hard to believe that the *Roads for Prosperity* programme announced then included a hugely expanded trunk-roads programme.

Its more detailed follow up *Trunk Roads England into the 90's* elaborated the theme. The individual proposals were estimated to affect 161 SSSIs directly or indirectly plus many local wildlife sites. In addition, there was the enormous impact of the total infrastructure encouraging more and longer movements by road and all that goes with it. This level of potential damage was unacceptable.

Since 1991 the roads programme has been trimmed back and back. Unusually, perhaps for many of us (!), the Treasury proved a sound ally in demanding public expenditure cuts in response to the recession in the early 1990s. When the new Government came in last May, there were still 118 road proposals being considered but rather than 161 SSSIs directly or indirectly under threat, we were down to about 44.

Other changes occurred in the 1990s notably the creation of the Highways Agency from the old Department of Transport. Whilst inevitably the Agency was focused on more roads, it also sought, with some successfully, to improve its own environmental standards and to take mitigation measures sensitively. Increasingly it will need to become more imaginative in seeking solutions to traffic congestion and demand without building more roads. Thus today the position is radically different from only 9 years ago. Something the environmental movement has long wanted seems within reach.

## DIRECT IMPACT OF ROADS ON WILDLIFE

What about some of the direct effect of roads and wildlife? In 1992 English Nature published a major report, *Roads and Nature Conservation – guidance on impacts, mitigation and enhancement*. Direct damage by road construction to habitats of value to nature conservation is obvious. Less obvious is the unseen degradation of biodiversity caused by isolation of habitats and sometimes the increased traffic itself. The 1992 report summarised the information on wildlife kills from a range of sources. English Nature Research Report No. 178 in 1995 summarised the issues on wildlife casualties more fully. Otters *Lutra lutra* have been successfully re-introduced into parts of England and it amazes me the number of these scarce animals which are killed on roads. For example no less than 10 in the Ouse Basin between 1992 and 1995. That tells us something about how fragmented wildlife areas have become.

Road construction is a major consumer of non-renewable resources. Of course sometimes wildlife benefits from sand and gravel extraction. There are several important wildlife sites in England resulting from these activities. There are also much less desirable impacts. We need to develop better policies for wise use of these non-renewable resources and for their recycling.

Road transport is a major contributor to air pollution. We all know that from our day to day experience in towns and cities but the wider effects are less obvious.

## INDIRECT EFFECTS

There are indirect effects like artificial lighting and noise disturbance. We recently published a research report on this subject and welcome the work by the Highways Agency to try and ameliorate the effects. Secondary development near new roads is again something familiar to us all. The CPREs report in 1992 entitled, *Concrete and Tyres – local development effects of major roads* is the best summary. The generation of more traffic is also well known.

The recent report by the Standing Advisory Committee on Trunk Road Assessment (SACTRA) emphasised that better access to an area to help

economic development can just as easily have the reverse effect.
Commuting distances have grown over the last 20 years; the average
distance of travel to work has grown by about 60%.

## NON TRUNK ROADS

Much of English Nature's work is directed at the Trunk Road
Programme. There is a substantial and much smaller scale non-trunk road
programme. The issues are identical but the local programme is not strategic
and can be driven hard by local political interest. The Wildlife Trust's
Report on their *Head on Collision* series illustrate the scale of the problem
well.

### Nature bites back

Just occasionally, very occasionally, natural forces take a road away.
The road to Spurn Point, Yorkshire, provides a graphic example. Can we
survive without it?

## ENGLISH NATURE'S POLICIES

English Nature sets out its policy position through Position Statements.
We published one on roads and nature conservation in 1992 and reviewed it
last year. The main principles advocated are:

- Better strategic planning to minimise travel.

- Equal evaluation of alternative modes of transport.

- The requirement for strategic environmental assessment.

- The need for early consultation and good information to allow informed
  decisions.

- Vigorous opposition to proposals that adversely affect SSSIs
  irreversibly.

- The need for effective mitigation and compensation.

In our role of statutory consultee, we demand consultation at the earliest
practical point in the road planning process. The Highways Act of 1980
stipulates that English Nature must be consulted on any road scheme passing
within a 100m of an SSSI. Of course, nowadays every major road proposal
is the subject of an environmental impact assessment. But to minimise or

prevent specific damage by individual roads, early input at the scoping stage is essential.

Whenever major work begins, we recommend the appointment of a full time clerk of works to oversee operations affecting wildlife interest during construction.

## THE ROADS REVIEW

For the current roads review Government has introduced a new process. Not only is it consulting more widely but also developing a new evaluation system. From our point of view, to have a specific evaluation dealing with nature conservation, completed by English Nature, which is set alongside all the major determinants is a significant step forward. No longer are we driven solely by items which can, allegedly, be costed, no matter how much many of us may believe that some costings are spurious.

In responding to the consultation document *What Role of Trunk Roads in England*, we recommended five general principles:

- Use traffic forecasts which take full account of targets to reduce carbon dioxide emissions and prevent air pollution.

- Make a strong presumption against new trunk roads off existing road lines.

- Subject whole routes to strategic environmental impact analysis.

- Set clear goals to sustain the total interest of wildlife and natural features.

- Include full environmental costs within the costs of transport use.

The new appraisal framework for road transport investment should ensure a greater weight being attached to environmental matters. However, we must recognise that the final decision is political. Ministers will always have a difficult task in weighing up the pros and cons of particular options. In the past we have lacked the confidence that the evaluation has really examined thoroughly all the potential impacts and options.

Decisions, therefore, need to be backed up with a clearer justification based upon the evidence which Ministers use to take difficult decisions. Furthermore, the decision on any individual road is certain to lead to both cries should of joy and anguish!

## INTEGRATED TRANSPORT POLICY

How do we develop a more integrated and environmentally sustainable transport policy? This must surely be Government's greatest challenge in the whole transport arena; the Deputy Prime Minister John Prescott, has emphasised his commitment to making real progress.

Think back to the early difficulties of bringing in lead free petrol, despite the minimal effects on most individual car users. If you then magnify the difficulty of that change by a 100, or perhaps even a 1000, you begin to recognise the size of the challenge. It will take 20 years of consistent direction, backed up with an ever-changing dynamic programme, to deliver a new regime. Altering the attitude of people is one of the most difficult. No Government will achieve much support for a transport policy which cripples employment, economic prosperity and individual freedom.

How are we to judge the success of an integrated transport policy? Let me give a few items which English Nature believes need to be done:

- Environmental standards and targets for the transport sector.

- Provide more investment in public transport to create more realistic affordable and attractive alternatives to the car.

- Use economic instruments to protect the environment and dampen the demand for travel by car.

- Use of the planning system better to reduce the need to travel.

These are huge and challenging issues to  provide strategic direction which will deliver what we need. Those of us involved in environmental conservation need to work hard on our constituency of direct public support to help bring about these changes.

They involve changes of lifestyle, attitude and opportunity, some of which will be highly unpalatable to large sections of the population. For example one only has to look at the trend, and popularity, for large stores built on the edges of our towns and cities.

- Are we going to stop having such stores in future? If so, at what rate should we expect them to decline?

- How do we need to organise future retail patterns to give people the opportunities and choices they have come to expect while at the same time doing better for the environment?

## CONCLUSIONS

Most of what I have said about roads is negative. In some ways this is unfair and I can imagine that those who build roads become fed-up with the criticism they receive. The brutal fact is that road development is not good for the environment. No matter how sensitively it is done, some erosion of what I term the 'quality of biodiversity' is almost inevitable. English Nature's geological responsibilities do sometimes benefit from road construction. Just as there are many old railway cuttings which produce fine rock exposures, the same is true of some roads. Admittedly visiting some of these sites is somewhat hazardous!

There are numerous roads in this country and they are here to stay. It is important therefore that we minimise their effects and make what gains we can. We have been involved with the Cornwall Wildlife Trust doing a wildlife audit of roadside verges in Cornwall. This audit has resulted in management guidelines to provide best practice examples for use by transport engineers and estate managers.

Barn owls *Tyto alba,* hunt beside roads and the number killed each year is depressingly large. It is often very difficult to tell how far wildlife kills on roads are significant in population terms but they are no less distressing. Techniques, of course, are used to limit wildlife casualties with badger holes, toad tunnels and the like. The expectation must be that more care is taken and that expense must be borne by the road developers.

Let me finish with one brief anecdote. In 1984, when the Nature Conservancy Council employed me as their Senior Ornithologist, I appeared at an inquiry at a proposed new road on the west side of Radipole Lake in Dorset.

We succeeded in altering the route of the road with well co-ordinated and organised opposition between ourselves, voluntary conservation organisations and a number of other local groups. The road was built but on a line which avoided the SSSI but not the local authority owned golf course. When I appeared at the enquiry the headlines in the local newspaper were not dominated by the road alignments. Part of my evidence about Cetti's warbler *Cettia cetti*, was featured. For those of you unfamiliar with this small bird, it colonised the southeast of England in the 1960s and is a skulking bird of thick damp vegetation beside reedbeds and rivers. The inspector was immensely interested in its sex life. The males are vigorous and spend their time looking after a substantial harem. This, rather than the road issues, made the headlines, and it illustrates how difficult it is to have the serious environmental issues debated and evaluated.

We need a better-informed public to recognise a wide range of environmental and social issues. I wonder what will happen in the future! As we enter the new Millennium, we shall see whether the delivery of an integrated transport strategy becomes a reality.

## The future for roads

Since 1998, we have gone forward and backwards on road issues. The ten year transport plan published in 2000, sets out the path for integrated transport but there has been a U-turn on road policy, back to the 1997 position. The political heat of the motoring lobby has largely caused the government to change its mind and build more roads. It has further encouraged road transport by lowering the cost of motoring, without any parallel change by lowering the cost of public transport. One fears that future budgets will further lower motoring costs and produce more roads and traffic.

# 5

# Balancing environment and transport

## J M DICKSON-SIMPSON

Few people are inclined to resist calls for environmental sensitivity. It is one of those worthy objectives that invite support from anyone with a conscience – akin to being in favour of safety.

But there is always a price at which idealism becomes tempered by pragmatism. Plans by the European parliament for a reduction in sulphur in diesel fuel to less than a thousandth of a per cent (a fiftieth of the present limit, which is a quarter of that a year ago) mean a £33,000 million bill for reformulating oil refineries. With exhaust emission standards even tighter than at present intended, the cost soars to over £90,000 million. Current exhaust emission constraints on diesel buses and lorries have already put £2,100 on the price of each vehicle. Future limits will add another £3,000 and to add catalytic oxidizing of residual soot particles will cost up to £5,000 a vehicle. These estimates are without inflation, and for Europe's lorries alone imply an extra cost of about £2,000 million a year in ten years' time.

Closer to people's pockets, fuel duty (roughly 60% of the retail price and 70% of the price businesses pay) would on pre-budget intentions double by 2005. Yet already people living in rural areas are complaining. Fuel duty raises £23,000 million a year. How much will be used to relieve congestion and subsidise adequate public transport is somewhat indeterminate.

All these massive items of transport cost are claimed to be in the cause of clean air. Are they worth the result? That is the question beginning to be asked. Yet the environmental crusade for even tighter limits stays rampant. It is time for a pause to take stock.

The emissions from new diesel engines are already a fifth of what they were in 1990. By 2002 they will be a seventh of the 1990 figures (see Figure 1). But it takes several years for a clean-up to spread to all the commercial vehicles in use. And further research to assess the true effect of the various kinds of pollutant and which ones ought to be targeted has only just begun.

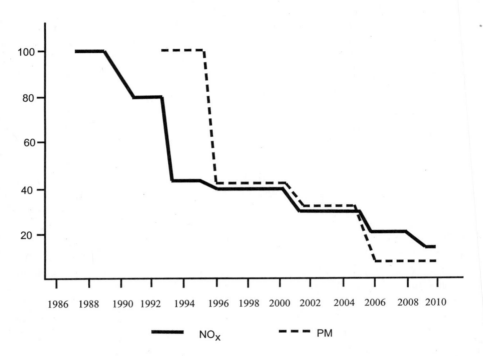

**Figure 1.** Evolution of exhaust emission standards for heavy-duty engines (simplified representation)

It is surely worth waiting to see the benefits of recent and near-future improvements before demanding more at increasingly higher cost. An era of diminishing returns has been entered. Despite the attention focused on diesel exhaust, the content of all its naughty pollutants is tiny – only 0.12% of the total volume of exhaust gas.

There are other facts to lend perspective. Out of 25,400,000 vehicles licensed in Britain, 420,000 – therefore, 1.65% – carry goods and gross above 3.5 tonnes.   Buses and coaches that seat more than eight people number 74,000, or 0.3%.

The number of lorries is going down, not up.   In 1985, 485,000 commercial vehicles above 3.5 tonnes gross moved 1,367 million tonnes of freight.  Ten years later, 421,000 moved 1,609 million tonnes. They are being worked harder and carry more per trip. The average yearly tonnage per commercial vehicle has risen by 35%. The ton-mileage – the work done – per vehicle has soared by 67%. The trend to fewer lorries will be reinforced as a result of the maximum gross weight being raised from 38 to 44 tonnes (with less road damage because there is another set of wheels to spread the load). Even on conservative estimates, the number of lorries should decline by about 4,000. Why 44 tonnes?  Because ships land containers weighing up to 30 tonnes, and a lorry to carry that weighs at least 14 tonnes.

**Figure 2.** Lorry with Roll-on/Roll-off body.

Restrictions on weights and lengths are the result of political compromises, not practical realities – not starting from the question 'what is expected to be carried'. Outstandingly efficient use of time comes from quickly detachable, roll-on/roll-off, bodies for lorries (Figure 2). Four at a time can be taken overnight to an out-of-town depot and then picked up by small trucks to nip about town. But the optimum nominal length of a compact demountable body to suit standardised packaging is 4 metres. They cannot be used, because Europe's Council of Ministers has decided that the total length of bodies on a lorry and trailer cannot be more than 15.65 metres, which is not a multiple of four. Political concern for the environment gets somewhat confused.

With the modem lorries, in recent years, the exhaust emissions have been cut by 80%, their noise has been halved, they have air springs to reduce vibration, special mudguarding to quell spray, collision absorbing guards all round, disc brakes that still stop if they are red hot. British lorries have been available with skid proofing since 1968 and it is now a legal requirement.

Their fuel consumption, and therefore carbon dioxide, has been cut by more than a third in 25 years.

There is a tendency for statisticians to forecast the future by extrapolating historic trends. This can be misleading. For example, one oft-quoted prediction is that the number of vehicles on the road will rise to 50 million. But that would be approaching one car for every person in the country, including children. Furthermore, the number of vehicles registered does not equate with the number actually on the roads at any given time. Suppose that environmental pressures persuaded people to buy electrically powered runabouts. That would mean second cars for many – a statistical scare, but on the whole only one would be used at a time.

The path to environmental heaven is strewn with paradoxes. Small commercial vehicles could replace big ones, but then between five and 18 times as many vehicles would be needed (depending on what was regarded as 'small').

Motorways can be made more attractive by lining them with trees. That not only can put more wild life exposed to danger but also can risk human life (there is hardly anything worse than colliding with a tree).

**Table 1** Share of Britain's goods movement. (Road movements include feeder-service for rail, air and water freight)

|               | % of freight | % of ton-miles freight activity |
|---------------|--------------|----------------------------------|
| Road          | 81.4         | 66.0                             |
| Rail          | 4.9          | 7.1                              |
| Air           | 0.1          | 2.0                              |
| Sea and canal | 6.7          | 20.2                             |
| Pipeline      | 6.9          | 4.7                              |

It is quite feasible for the amount of freight carried by rail to increase by 30% over the next couple of years. As, however, the proportion of freight on rail is hardly more than 6% at present, the resultant net decline in road freight would be only 2.7% – and on main roads (Table 1). The rail terminals still have to be served by road, and rural and urban traffic would be undiminished. Noise and spray can be halved simultaneously with open-texture road surfaces, but they cost about 10% more, so government has resisted them, until the ten-year transport plan takes effect.

Regional planning could reverse shopping patterns from outlying supermarkets to local communities. That would reduce car traffic but increase commercial-vehicle traffic.

It is fashionable to regard road transport as a nuisance rather than an asset. Roads are great conveniences, extremely well utilised. They are the routes to employment for many and raise the value of land they serve; there's the rub. Improved access to road transport raises the value of land without any effort by the landowners. Instead of thinking in terms of taxing the use of roads, it would be more logical to tax the enhanced value of the land they serve.

# 6

# Treatment of biodiversity issues in road Environmental Impact Assessments

H BYRON & W SHEATE

While ecological assessment has always been an integral component of Environmental Impact Assessment (EIA), biodiversity is a much more recent concept. This paper examines the extent to which impacts of roads on biodiversity are addressed in EIAs and then seeks to explore how consideration of biodiversity through the EIA process can be improved. Wildlife impacts have been an important catalyst for much of the controversy over recent road building programmes, sometimes only being identified late in the day. This suggests that even traditional ecological assessments are not always as comprehensive as they might be. The development of an approach for considering systematic biodiversity issues in the road scheme EIA process would, it is argued, give biodiversity its appropriate status in road and transport planning.

Over the last decade, the term biodiversity has come into widespread use; however, despite acceptance of the term there is no consensus definition. DeLong Jnr. (1996) and Takacs (1997) reviews over 85 different definitions. These include – *"For the purpose of the Global Biodiversity Assessment, biodiversity is defined as the total diversity and variability of living things and of the systems of which they are part. This covers the total range of variation in and variability among systems and organisms, at the bioregional, landscape, ecosystem and habitat levels, at the various organismal levels down to species, populations and individuals, and at the level of the population and genes. It also covers the complex sets of structural and functional relationships within and between these different levels of organisation, including human action, and their origins and evolution in space and time."* (UNEP Global Biodiversity Assessment, 1995)

The perceived strengths and weaknesses of the broad concept of biodiversity and the distinctions between 'biodiversity' and 'wilderness' and 'nature' are the subject of much discussion (Takacs, 1997). To an extent, biodiversity is a political concept which conservation biologists have created and emphasised in an attempt to shift the focus of conservation from endangered species towards a more integrated ecosystem approach.

Biodiversity was placed firmly on the international agenda on signature of the Convention on Biodiversity, by representatives from over 150 countries (including the UK), at the 1992 UNEP Earth Summit in Rio de Janeiro (UNCED, 1992). The Convention places various obligations on signatories. These obligations include requirements to produce a national biodiversity plan and to ensure that strategies for conserving biodiversity are integrated into sectoral policies, and to introduce procedures and arrangements to ensure that impacts of projects, programmes and policies on biodiversity are assessed. The relevant provisions of the Convention are set out below.

### Article 6
"Each Contracting Party shall, in accordance with its particular
conditions and capabilities:

(a) Develop national strategies, plans or programmes for
the conservation and sustainable use of biological diversity
or adapt for this purpose existing strategies, plans or programmes
which shall reflect, inter alia, the measures set out in this
Convention relevant to the Contracting Party concerned;
and
(b) Integrate, as far as possible and as appropriate,
the conservation and sustainable use of biological diversity
into relevant sectoral or cross-sectoral
plans, programmes and policies."

### Article 14
"Each Contracting Party, as far as possible and as appropriate, shall:

(a) Introduce appropriate procedures requiring environmental
impact assessment of its proposed projects that are likely
to have significant adverse effects on biological diversity
with a view to avoiding or minimizing such effects and,
where appropriate, allow for public participation
in such procedures;

(b) Introduce appropriate arrangements to ensure that the
environmental consequences of its programmes and policies
that are likely to have significant adverse impacts on
biological diversity are duly taken into account; ...."

## IMPLEMENTATION OF BIODIVERSITY OBLIGATIONS

### The UK

In the UK, implementation of the obligations of the Convention has focused on the requirement to produce and implement a national biodiversity plan. In 1993, a group of six UK voluntary conservation bodies produced the discussion document *Biodiversity Challenge* (Wynne *et al.*, 1993, 1995) to aid the government in the production of a national plan. The *UK Action Plan on Biodiversity* was published in January 1994 (HM Government, 1994). This plan set out the broad strategy and targets for conserving and enhancing wild species and wildlife habitats for the next 20 years, and established a UK Steering Group (now the UK Biodiversity Group) to oversee certain tasks, including developing costed targets for key species and habitats. The Steering Group's findings, published in December 1995 (HM Government, 1995) and endorsed by the government in May 1996 (HM Government, 1996), make detailed proposals for a large number of species and habitats which require urgent conservation action. Costed action plans (Species Action Plans or Species Statements) have now been produced for over 459 key species and 38 key habitats (Habitat Action Plans (HAPs)), and 37 habitat statements for the broad UK habitat types (Habitat Statements (HSs)) (HM Government, 1995; English Nature, 1998a,b, 1999a,b,c,d). Local Biodiversity Action Plans (LBAPs) are seen as the mechanism for ensuring that the national strategy is translated into effective action at the local level. Guidance for the development of LBAPs has been issued recently (UK Local Issues Advisory Group, 1997; Scottish Biodiversity Group, 1997; Countryside Council for Wales, 1997; London Wildlife Trust, 1998). At present, implementation of the action plans (national and local) is at a preliminary stage.

Although UK production of the National Biodiversity Plan and supporting action plans is at a relatively advanced stage, there is no obvious implementation of biodiversity obligations in the UK via integration (pursuant to article 6) e.g. practical incorporation or guidance for incorporating biodiversity considerations in to EIA, Strategic Environmental Assessment (SEA), policy appraisal, etc (as mentioned in article 14).

The Countryside Commission, English Nature, English Heritage and the Environment Agency have proposed a new characterisation-based approach to assessing environmental capital. The approach is explained in detail in the recent publication *What matters and why? environmental capital: A new approach – a provisional guide* (CAG Consultants and Land Use Consultants, 1997).

This new methodology aims, *"...to stand back from environmental things or features, and consider the environmental functions they perform, or the services they provide for human well being"*. Having decided what area, feature or group of features is to be studied (which depend on the purpose of the assessment) the new approach asks the following.

- What are the *characteristics* or *attributes* of this place or object(s) which matter for sustainability?

- *How important* is each of these, to whom, and for what reasons or purposes?

- What (if anything) could replace or *substitute for* each of these benefits?

- On current trends do we expect to have *enough* of each of them?

The approach uses the answers to decide what kinds of management action are needed to protect and/or enhance each of the attributes.

The same environmental capital assessment process is intended to assess different kinds of environmental interests including biodiversity, landscape, heritage, etc, and provide useful input to decision processes. This approach has been developed to link in with other management tools and processes including EIA and SEA. In relation to EIA the report states *"...the new approach has the benefits of bringing effects beyond the development site into the analysis and thus allowing comparison of different options – two areas often weak in traditional environmental assessment"*. Hence, the proposed methodology could be used to strengthen EIA in these two areas generally (including the biodiversity aspects) in the future.

## Implementation via the EIA process:
## International guidance on biodiversity in EIAs

Although the need to amend existing EIA practice to encompass biodiversity impacts has been acknowledged, at present there is little guidance to help this process. Guidance has, however, been issued outside the UK by the US Council on Environmental Quality (CEQ)(CEQ, 1993), the Canadian Environmental Assessment Agency (CEAA)(CEAA, 1996), the World Bank (1997, 2000) a framework approach to biodiversity has been discussed by the International Association for Impact Assessment (IAIA) (Sadler, 1996), workshops on biodiversity impact assessment held at the 1998, 1999 & 2000 annual IAIA meetings and the IUCN is part way through a programme of work for 'Addressing Biodiversity Impact Assessment' (IUCN web page, 1998). The US guidance is discussed further below as one example of how existing EIA practice can be expanded to encompass biodiversity impacts.

## US CEQ guidance

In the USA under the National Environmental Policy Act 1969 (NEPA) and the executive orders and regulations of the CEQ, EIA is mandatory for every recommendation or report on proposals for legislation and other major federal actions significantly affecting the quality of the human environment.

In 1993 the CEQ published a report entitled *Incorporating biodiversity considerations into environmental impact analysis under the National Environmental Policy Act* (CEQ, 1993). This report summarises emerging biodiversity concepts and practices and how they may be applied to NEPA analyses, but it is not formal CEQ guidance. The report discusses what is meant by the term biodiversity, sets out the main objectives of biodiversity conservation and management together with key principles for conserving biodiversity (including the key concept of the ecosystem approach), and explains how biodiversity can be incorporated into NEPA analysis. Several of the key issues from the report are outlined below.

In particular the report stresses the importance it places on the incorporation of biodiversity issues into the NEPA process stating: *"The extent to which biodiversity is considered in future NEPA analyses of federal actions will strongly affect whether biodiversity is adequately protected in the coming decades. It is critical that federal agencies understand and take into account general principles of biodiversity conservation in their decision-making"* and that *"Successful implementation of the principles of biodiversity management requires that they be effectively integrated into the NEPA process".*

The guidance sees biodiversity as comprising:

- *Regional ecosystem diversity:* The pattern of local ecosystems across the landscape, sometimes referred to as 'landscape diversity' or 'large ecosystem diversity'.

- *Local ecosystem diversity:* The diversity of all living and non-living components within a given area and their interrelationships. Ecosystems are the critical biological/ecological operating units in nature. A related term is 'community diversity' which refers to the variety of unique assemblages of plants and animals (communities). Individual species and plant communities exist as elements of local ecosystems, linked by processes such as succession and predation.

- *Species diversity:* The variety of individual species, including animals, plants, fungi and micro-organisms.

- *Genetic diversity:* Variation within species. Genetic diversity enables species to survive in a variety of different environments, and allows them to evolve in response to changing environmental conditions.

In relation to the basic goals of biodiversity conservation and management the report states:
*"The basic goal of biodiversity conservation is to maintain naturally occurring ecosystems, communities, and native species."* and *"The basic goals when considering biodiversity management are to identify and locate activities in less sensitive areas, to minimise impacts where possible, and to restore lost diversity where practical"*.

As noted above, the guidance considers the ecosystem approach to be a fundamental concept. This approach includes the analysis of both the elements and the interrelationships involved in maintaining ecological integrity and uses a local-to-regional perspective that considers impacts at the appropriate scale within the context of the whole system. The guidance advocates that even at the project-specific or site-specific level, analyses should extend to the regional ecosystem scale to consider adequately impacts on biodiversity (CEQ, 1993: CEQ web pages, 1998).

The guidance sees implementation of an ecosystem framework including (1) selection of the appropriate scales of analysis, and (2) establishment of goals and objectives for the protection of biodiversity, based on (3) an adequate information base. In relation to this the guidance discusses the issue of scale which it views as a central issue in the ecosystem approach. The guidance sees that the appropriate boundary for a NEPA analysis as one that ensures adequate consideration of all resources that are potentially subject to non-trivial impacts and explains that for some resources, that boundary can be very large.

Examples are given of the long-range atmospheric transport of nutrients and contaminants into water bodies such as the Great Lakes and Chesapeake Bay transcending even the boundaries of their vast watersheds, and at the other end of the spectrum, how significant contributions to biodiversity protection can be made by identifying and avoiding small sensitive areas, such as rare plant communities. The guidance suggests that determining relevant boundaries for assessment should be guided by informed judgement, based on the resources potentially affected by an action and its predicted impacts.

To conclude, the CEQ guidance is valuable and provides useful guidance on the ecosystem approach and its application. However, the guidance does not detail the practicalities of carrying out an assessment of biodiversity within an EIA using the ecosystem approach. In part this could be a consequence of the wide scope of NEPA (which applies to recommendations or reports on proposals for legislation and other major federal actions) i.e. the guidance may be broad and general in order to be applicable to the full range of levels of assessments carried out pursuant to NEPA.

## CURRENT ROAD EIA OBLIGATIONS IN THE UK

### EIA

Various definitions have been proposed for EIA including *"... a process which seeks to identify, predict, evaluate, mitigate and communicate the environmental consequences of a proposal"* (Munn, 1979) and *"... an assessment of the impact of a planned activity on the environment"* (UNECE, 1991).

EIA has been a legal requirement for certain projects in the UK since EC Directive 85/337 (Council Directive on the Assessment of the Effects of Certain Public and Private Projects on the Environment) (the EIA Directive) (CEC, 1985) was implemented in 1988. The EIA Directive was subsequently amended by EC Directive 97/11/EC (the EIA Amendment Directive) (CEC, 1997) The EIA Directive requires mandatory EIAs for 'Construction of motorways, express roads...'. An EIA may be required for other types of road project depending on the characteristics of the particular project. The provisions of the EIA Directive relating to trunk roads are implemented in the UK by the Highways (Assessment of Environmental Effects) Regulations 1999, Part III of the Environmental Impact Assessment (Scotland) Regulations 1999, and the Roads (Environmental Impact Assessment) Regulations (Northern Ireland) 1999, and the Town and Country Planning (Environmental Impact Assessment) (England & Wales) Regulations 1999, Part II of the Environmental Impact Assessment (Scotland) Regulations 1999, and the Planning (Environmental Impact Assessment) Regulations (Northern Ireland) 1999, for local highway schemes.

The EIA Directive specifies that − *"... the EIA will identify, describe and assess ...the direct and indirect effects of a project on the following factors:*

- *human beings, fauna and flora,*
- *soil, water, air, climate and the landscape,*
- *the interaction between the factors mentioned in the first and second indents,*
- *material assets and cultural heritage."*

As ecology is the scientific study of the interactions between living organisms and their environment the wording of the EIA Directive clearly requires the ecological implications of proposed road projects to be studied. (Obviously the wording of the Directive also requires the other environmental impacts of proposed developments to be considered e.g. impacts on air quality, cultural heritage, water quality and drainage, and geology and soils, and the complete EIA will address all these issues.)

Although its preamble does refer to the 'maintenance of the diversity of species', the EIA Directive (CEC, 1985) does not specifically refer to impacts on biodiversity (it was agreed in 1985 before the Biodiversity Convention, and the EIA Amendment Directive 97/11/EC (CEC, 1996) does not change this). But such impacts clearly fall within the scope of the EIA Directive and should therefore be considered in EIAs. Indeed, as biodiversity has now become such a major issue it seems likely that if the EIA Directive was being negotiated now it would specifically refer to impacts on biodiversity.

## Ecological and biodiversity impacts

Ecology looks at the relationship between organisms and their environments at three levels: species, communities and ecosystems. Biodiversity is concerned with diversity at the levels of species (diversity both within and between species) and ecosystems. As ecology and biodiversity are not distinct concepts, the distinction between ecological impacts and biodiversity impacts in practice is not clear. Impacts of a proposed road development could be:

- Ecological impacts = any impacts on species, communities and ecosystems.
- Biodiversity impacts = any impacts affecting the genetic variation within a species or the diversity of species and ecosystems.

Biodiversity impacts could perhaps be considered as the category of ecological impacts concerned with the reduction of diversity. However, there appears little recognition of this distinction in the EIA literature which appears to assume that ecological impacts and impacts on biodiversity are synonymous.

## Ecological assessment guidance

Current UK guidance on the preparation of the ecological components of road EIAs is contained in Volume 11 of the Design Manual for Roads and Bridges (Volume 11) (DOT, 1993). This guidance superseded the Manual of Environmental Appraisal (DOT, 1983) and Departmental Standards TD 12/83 and HD 18/88 in relation to road schemes which enter the Roads Programme after June 1993.

Volume 11 gives guidance on the level of ecological assessment required as a road scheme develops – *"Ecological assessments should become increasingly detailed as a scheme develops and in accordance with the ecological importance of the area affected by the scheme"*. At the stage at which a formal EIA is required Volume 11 states: *"The objective at this stage is to identify any significant nature conservation impacts likely to arise from construction of the preferred route, and to identify the location, type and importance of all areas of significant nature conservation interests that may be affected"*.

Volume 11 does not specifically refer to biodiversity or the UK obligations under the Convention, although it notes that: the objectives for nature conservation are:

- The maintenance of the diversity and character of the countryside.

- The maintenance of viable populations of wildlife species, throughout their traditional ranges, and the improvement of the status of rare and vulnerable species.

Volume 11 identifies a number of main potential impacts on nature conservation:

- Direct loss of habitats through land take.
- Severance.
- Wildlife casualties.

- Disruption to local hydrology.
- Pollution of local watercourses by road run off.
- Road structures causing problems for certain birds and mammals.
- Effects of road lighting.
- Effects of air pollution from vehicle emissions.
- Road spray.
- Disturbance during construction.

In relation to areas of high nature conservation value Volume 11 emphasises protected species and designated sites, although it does discuss the need to consider the nature conservation value of other areas and sets out a list of features considered indicative of potential nature conservation value e.g. river or stream valleys, areas of permanent pasture and herb rich meadow, and which if present should be considered. DOE (1995) guidance on the preparation of planning EISs, even though two years after Volume 11, still makes no mention of biodiversity, concentrating as before on ecological impacts.

## CURRENT STATUS OF BIODIVERSITY IN UK ROAD EIAS

### 1993 EIS review

A review of 37 road Environmental Impact Statements (EISs) (Treweek et al., 1993) found a number of shortcomings in the assessment of potential ecological impacts arising from the development of new roads. This review did not address the issue of impacts on biodiversity, but looked more widely at potential ecological impacts of roads and the quality of EISs in this respect. The review found that in some respects the ecological components of EISs failed to comply with the requirements of the EIA Directive. In particular it found:

- Few EISs stated the size of the proposed road development (whether in terms of length or land take of major wildlife habitats).
- There was a widespread lack of the ecological data necessary to identify and assess the main effects which the development will have on the environment.  This was primarily due to inadequate ecological surveys. In many cases it appeared that no surveys had been carried out, and

where surveys had been carried out these were largely descriptive and focused almost entirely on the distribution of broad habitat types. The review commented that the lack of any information about the distributions of species and their populations makes it impossible to assess the likely implications of the proposed developments on those species on the basis of the information in the EIS.

- 13.5% of the EISs completely omitted potential ecological impacts.

- Only 9% of EISs quantified impacts.

- Generally EISs only discussed direct impacts such as habitat loss.

- There was a general lack of baseline data, which the review noted is currently unlikely to be remedied because of the general lack of commitment to follow up monitoring.

- No assessment of possible cumulative effects in terms of the local, regional and national status of habitats and species.

- Proposed mitigation measures were vague and did not relate to specific impacts. No estimates of their likely success were indicated and in very few cases were detailed prescriptions given.

- The mitigation emphasis was almost entirely on aesthetic compensation (tree planting and landscaping), rather than measures to mitigate ecological damage, such as loss of habitats and species or alterations in their status or stability.

Overall the review concluded that many EISs failed to provide the data necessary to predict potential ecological impacts and very few attempted to quantify these impacts.

**1997-8 EIS review**

The current treatment of ecological/biodiversity impacts in road EIAs is being investigated by a further review of recent road EISs (EISs dated June 1993 or later) to follow up the previous work (outlined above) and in particular to see how biodiversity impacts are currently treated in road EISs. Forty EISs have been reviewed (Byron et al., 2000) and this has revealed:

- Assessment of potential ecological effects improved in some respects, but many of the weaknesses found in the 1993 study still appear.

- None of the EISs specifically referred to potential impacts on biodiversity.

- The statements focus heavily on possible impacts to designated sites and protected species, with 40% of the EISs referring to Sites of Special Scientific Interest (SSSIs), 85% referring to sites designated at a local or county level and 80% referring to protected species.

- While the majority of EISs (95%) did identify potential impacts on other sites and species, in general these were not addressed in detail.

- Potential ecological impacts were discussed in all of the statements, but the majority of statements still only made reference to more direct impacts such as habitat loss and habitat severance.

- 40% of the statements made some attempt to quantify the area of habitat to be lost to the scheme, but only one EIS attempted to quantify any other potential impacts.

The types of ecological impact referred to included:

- Direct habitat loss (100%)

- Habitat severance (48%)

- Habitat fragmentation (10%)

- Disturbance (53%)

- Water pollution (28%)

- Potential effects of lighting (7%)

- Road casualties (15%).

The lack of discussion of the potential impact of habitat fragmentation is particularly concerning as this is thought to be one of the principal causes of loss of biodiversity.

- Ecological surveys were found to be inadequate:
  - Although new surveys were carried out in 88% of the EISs the vast majority of surveys focused almost entirely on the distribution of broad habitat types and were largely descriptive
  - Only 38% of the EISs contained any quantitative data about the distributions of species.

- Only one EIS referred to possible cumulative effects in terms of the local, regional and national status of habitats and species.

- Mitigation measures were addressed generally in all of the statements, but only eleven EISs gave detailed prescriptions for suggested mitigation measures.

- As with the 1993 study, the emphasis was on aesthetics and compensation measures which might improve the appearance of road verges (tree planting and habitat creation). However, tunnels for animal movements were discussed in 43% of EISs and pollution traps in 65%.

This recent work confirms that impacts of roads on biodiversity are not being satisfactorily addressed in road EIAs. Particularly that there is a lack of:

- A broader ecosystem approach to provide a context for the assessment of impacts.

- Consideration of impacts on non-protected species.

- Consideration of impacts on non-protected areas.

- Consideration of a full range of impacts (especially more indirect impacts).

- Consideration of cumulative impacts.

## A WAY FORWARD FOR ROAD EIAS –
## THE BIODIVERSITY ASSESSMENT METHODOLOGY

### Background

The impetus for this project were concerns that despite the UK biodiversity initiatives and the EIA process, biodiversity considerations are not being taken into account in sectoral decisions such as transport policy decisions and decisions on individual road schemes. To address these concerns in relation to transport decisions various UK organisations (principally RSPB, WWF, English Nature and the Wildlife Trusts) formed the Transport and Biodiversity Group (TBG). This research was partly sponsored by the TBG (with the Economic and Social Research Council) and was a 3 year project which was completed in mid 2000.

The aim of the project was to develop a method/procedure for the systematic treatment of biodiversity issues in the road scheme EIA process (the methodology). This method fits into the current ecological impact assessment process to strengthen EIAs in relation to biodiversity.

Potentially the methodology could aid the UK in meeting its objectives under the Biodiversity Convention to maintain naturally occurring ecosystems, communities, and native species,  by ensuring that road development proposals do not significantly reduce the variability among and within living species and organisms, including the ecosystems and communities of which they are part.

The methodology was developed particularly in relation to EIAs for new roads. This is partly because the project was originally  conceived during the recent era of the much expanded roads programme where there were various instances where the planning and decision-making process did not give priority to the protection of national sites of wildlife interest with the result that a number of SSSIs were irreversibly damaged by road building. In the current climate, despite the reduction of the roads programme, there will still be a programme of new road building and upgrades of existing roads (as part of integrated transport solutions for particular areas) for which the methodology will be necessary. In addition, the methodology could also be used to assess operational/management impacts of existing roads to ensure that these are not contrary to biodiversity objectives, could be adapted for use in EIAs of other types of project, and may also be applicable to other levels of decision-making e.g. policy, plans and programmes.

**Project work programme**

The project consisted of the following stages:

- General literature review.

- Review of EISs.

- Literature review of guidance for treatment of biodiversity in EIAs and methods of measuring biodiversity e.g. indicators.

- Consultation on measurement methods and approaches to the assessment of 'biodiversity value' with a range of thirty-one experts. These experts included TBG, The Highways Agency, The National Assembly of Wales, English Nature, Countryside Council for Wales, members of Infra Eco Network Europe (IENE), members of the international impact assessment community, environmental consultants, local authority planners, ecologists and academic ecologists.

- Selection of measurement methods – this formed a key element of the project.  As biodiversity encompasses variability at various different levels there is no single straightforward method for assessing it. Measurement of biodiversity is discussed further below.

- Consultation – feedback and modification of the methodology as appropriate.

- Case study selection – based on selection criteria (outlined further below).

- Testing the methodology with tour case studies.

- Revision of the methodology in light of case studies.

- Production of guidance for implementation of the methodology – the final methodology comprises a key objective and set of guiding principles, plus guidance for how to address biodiversity impacts at each stage of the EIA, a biodiversity checklist intended for use as a final check to ensure that an EIA has considered all relevant biodiversity issues.

Guidance on the methodology has now been published in the form of a leaflet *Biodivesity impact – Biodiversity and Environment Impact Assessment: A New Approach* (RSPB *et al.*, 2000) and a full report *Biodiversity Impact – Biodiversity and Environmental Impact Assessment: A Good Practice Guide for Road Schemes* (Byron, 2000).

Figure 1 shows an overview of the project.

**Measurement methods**

As noted above, as biodiversity encompasses variability at different levels there is no single simple and accepted method for measuring it. Measurement has traditionally focused on the species level and indeed measurement of species richness and/or abundance can give a very detailed picture of the biodiversity in a particular place, but it is time and cost intensive and requires detailed expert knowledge – one factor, perhaps, as to why it is so poorly addressed in EISs. If suitable indicators were available this could facilitate less cost intensive assessments of biodiversity for use in EIAs. However, to date there are no widely used indicators of biodiversity.

The current emphasis focuses on measurements of biodiversity most appropriate for the particular purpose for which the information is to be used. Hence, the project investigates a range of possible biodiversity measurement methods to determine which are the most appropriate /informative for strengthening EIAs of road projects.

Some of the options considered for measuring, and indicators of, biodiversity are shown in Figure 2.

## Case study criteria

Criteria on which the selection of the four case studies made are:

- A recent road project (ideally post 1993).
- Raising biodiversity issues i.e. some biodiversity constraints. The case studies investigated a range of different scales e.g. areas of international significance for biodiversity and areas of local significance.
- Preferably with habitats with a HAP/HS e.g. hedgerows or subject to the Habitats Directive.
- Preferably with species with a SAP or subject to the Habitats Directive.
- If UK ideally in an area for which a LBAP has been prepared.
- EIS available.
- Further data available.
- Range of project locations (in/outside the UK). Two of the cast studies were in England, one in Wales and the fourth in Canada.

## Key objective and background principles

The key objective of the methodology is to ensure that road schemes do not significantly reduce biodiversity at any of its levels and enhance biodiversity wherever possible.

Appropriate background principles to guide biodiversity assessments have been adapted from existing guidance on assessment of biodiversity impacts (CEQ, 1993, CEAA, 1996, IUCN, 1997 and Canter, 1997) and modified as a result of the consultations. The final guiding principles are shown in Figure 3.

## Interaction with the EIA process

An overview of how the methodology ties in to the EIA process is shown in Figure 4. This includes examples of some of the biodiversity considerations which will need to be addressed at different stages of the EIA process.

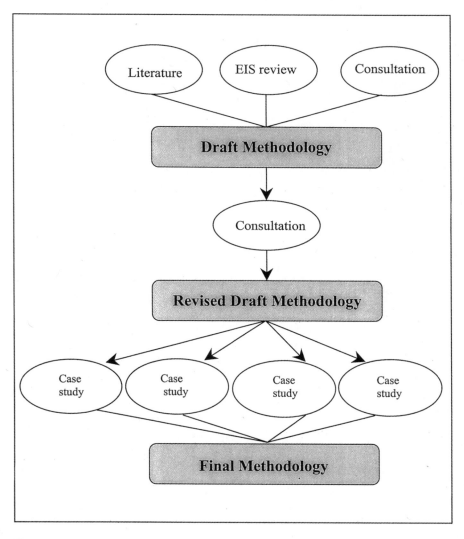

**Figure 1.** An overview of the project.

## MEASUREMENTS

### Habitat/ecosystem

-no, identity and distributions of habitats

-no, identity and distributions of land uses

-value of protected habitats (local, regional.....)

-assessment of biodiversity potential of non
    protected habitats

-relevant HAPs (national, regional and local)

### Species

-species richness

-species composition

-species abundance of key species

-species distribution of key species

-relevant BAPs (national, regional or local)

### Genes
-allelic diversity

## INDICATORS

### Habitat/ecosystem

-ecosystem diversity

-area in protected status

-% of area of "biodiversity value" not of protected
    status

-total % of natural/semi-natural vegetation cover

### Species

-species richness and composition

-family richness

-endemic species richness

-threatened species richness

-BAP (national, regional or local) species

### Genes

-presence of rare alleles

**Figure 2.** Some of the options considered for measuring, and indicators of, biodiversity.

Some of these considerations are explained in more detail below.

### Screening

What biodiversity considerations should trigger an EIA? For example does the project:

- Impact on a protected area?

- Impact on biological resources important for the conservation of biodiversity?

- Impact on attempts to protect ecosystems or promote the recovery of threatened species (HAPs and SAPs etc)?

### Scoping

The aim of this stage is to determine the scope of the EIA in terms of:

- The specific aspects of biodiversity to be addressed (the species, communities, and ecological processes which could be impacted by the project and their status (the receptors))

- Identification of potential impacts, in relation to cumulative effects identification of other projects/activities that should be considered.

- The spatial context of the EIA – identification of the local to regional study area and the spatial issues which need to be considered for each receptor.

- The time parameters of the assessment – identification of the time issues which need to be considered for each receptor, this is especially important where the project affects migratory species.

### Baseline studies

This stage aims to assess potential receptors and evaluate their sensitivity and/or biodiversity value. It will be a collation of existing data on biodiversity in the study area from a wide range of sources supplemented by new fieldwork.

### Impact prediction

The aim at this stage is to assess the magnitude of each potential impact on biodiversity and to quantify impacts as far as possible. The detail of prediction for impacts will vary. For example, direct land take/habitat loss is quantifiable, for habitat severance and habitat fragmentation the size of habitat lost/remaining is quantifiable and the ability of the remaining patches of habitat to support viable populations could be discussed, but would be

## Guiding Principles

✓Avoid impacts on biodiversity and create opportunities for enhancement of biodiversity wherever possible by route selection and scheme design. Where this is not possible identify the best practical mitigation and enhancement option to ensure that there is no significant loss of biodiversity. Compensation measures such as translocation should be viewed as a last resort.

✓Apply the precautionary principle to avoid irreversible losses of biodiversity. ie where an activity raises threats or harm to biodiversity precautionary measures should be taken even if certain cause and effect relationships are not scientifically established.

✓Widen existing EIA practice to an ecosystem perspective - ie consider the impacts of a road scheme on biodiversity and possible enhancements of biodiversity in the context of local and regional ecosystems, not just the immediate vicinity of the road.

✓Safeguard genetic resources by protecting the higher levels of biodiversity (ie individuals, populations, species, and communities, etc.) and the environmental processes which sustain them.

✓Consider the full range of impacts on biodiversity eg indirect and cumulative impacts not just the direct impacts such as species and habitat loss.

✓The study area of the scheme should reflect the impact type (eg indirect effects will often extend throughout a watershed) rather than taking a fixed width corridor approach.

✓Evaluate the impacts of a road scheme on biodiversity in local, regional, national, and, where relevant, international contexts ie an impact could be minor locally but significant at a national level eg where the locality has a very high proportion of a nationally rare biodiversity resource.

✓Retain the existing pattern and connectivity of habitats eg protect natural corridors and migration routes and avoid artificial barriers. Where existing habitat is fragmented implement measures eg tunnels, bridges to enhance connectivity.

✓Use buffers to protect important biodiversity areas wherever possible.

✓Maintain natural ecosystem processes in particular hydrology and water quality. Wherever possible use soft engineering solutions to minimise impacts on hydrology.

✓Strive to maintain/enhance natural structural and functional diversity eg ensure that the quality of habitats and communities is not diminished and wherever possible is enhanced by the road scheme.

✓Maintain/enhance rare and ecologically important species (key species) - ie protected species, SAP species, characteristic species for each habitat as loss of these may affect a large number of other species and can affect overall ecosystem structure and function.

✓Decisions on biodiversity should be based on full information and monitoring must be part of the EIA process. The results of monitoring should be available to allow evaluation of the accuracy of impact prediction and should be widely circulated to help improve future road scheme design and mitigation.

✓Implement on-going management plans for existing and newly created habitats and other mitigation, compensation and enhancement measures.

**Figure 3.** Guiding principles

hard to quantify unless extremely detailed data were available. In relation to road mortality, the species particularly at risk (e.g. reptiles and amphibians, birds of prey e.g. barn owls, mammals with large territories or habitual foraging routes e.g. otters, deer and badgers) can be identified and discussed, but it will be difficult to assess the effects road mortality will have on population viability as high mortality rates may represent severe impacts on local populations or alternatively may indicate that local populations are thriving. Indirect impacts, for example the effects of water borne pollution such as silt and chemicals in runoff and disturbance on biodiversity, are generally much more difficult to quantify, however these impacts must still be identified and discussed. Where quantification of impact magnitude is not possible impacts can be ranked e.g. on a scale of severe, serious, slight, none, uncertain, beneficial.

### Evaluation of impact significance

The 'significance' of an impact is a function of the magnitude of the environmental impact and the ecological status and sensitivity of the species and/or habitats affected. An impact may be significant because it affects a large area, or because it affects a small area which is sensitive and/or of high nature conservation value. Biodiversity impacts could be evaluated on a scale of very severe, severe, significant, minor and negligible using an evaluation matrix.

Evaluation of significance should also consider possible types of mitigation and should assess the potential impacts of the mitigation measures themselves so that the ultimate assessment of significance of impacts takes the effects of the mitigation measures into account. Where relevant, mitigation measures should be related to specific impacts and an assessment of the likelihood of success and details of implementation should be included.

### Decision-making

Biodiversity inputs throughout the EIA process would ensure that biodiversity is given proper weight in the decision making process.

### Monitoring

The current EIA process in the UK does not require monitoring to assess implementation of the mitigation or its effectiveness and so monitoring is usually neglected unless specifically required as a condition of consent. However, ecological good practice would include monitoring of residual impacts (which could be achieved by the methods used in the baseline

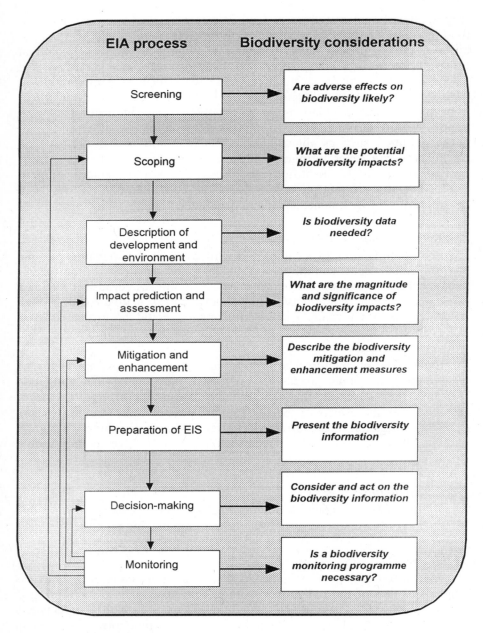

**Figure 4.** An overview of how the methodology ties in to the EIA process.

study).  The data from this monitoring could be used to provide information on residual impacts and also be used as cumulative data for future EIAs.

## Discussion

Despite the existence of the current ecological impact assessment process which implicitly addresses some biodiversity aspects, current EIAs of road projects are not incorporating the full range of biodiversity considerations.

Development of the methodology should strengthen consideration of biodiversity considerations in EIA and address certain specific weaknesses in current practice:

- Currently road EIAs are not addressing biodiversity considerations explicitly.  Some would argue that explicit treatment of such issues is unnecessary as EIAs are already dealing with biodiversity issues implicitly.  However it is argued that current implicit treatment of biodiversity issues is insufficient (from the results of the EIS reviews and for the reasons discussed further below) and that in any event EIAs should be explicitly addressing biodiversity.  In addition, the EIS review evidence shows that ecological assessments in road EIAs are still inadequate.

- It could be argued that current EISs do assess biodiversity implicitly in that they may contain habitat and species lists.  However, a species list, which looks at species richness alone, is only one method of measuring biodiversity (see measurement of biodiversity above) and does not address other fundamental aspects of biodiversity such as species abundance and distribution.  An EIA that only considers a species list is not thoroughly addressing biodiversity.

- The concept of biodiversity is essentially a political tool which has been adopted to provide a new focus and emphasis on the aims of nature conservation, especially on the priority given to the protection of species and habitats.  The central tenet of the Biodiversity Convention that conservation protection must be strengthened and the new approach emphasises not only the need to strengthen policy and decision-making specifically related to biodiversity conservation, but also the need to ensure that biodiversity considerations are incorporated in to policy and decision-making in other areas.  The latter is essential to ensure that achievements of specific biodiversity  policy/programmes, etc are not undermined by policies/programmes and decisions in other areas which are detrimental to biodiversity conservation. In light of this integration aim of the Biodiversity Convention it is essential that biodiversity considerations are dealt with explicitly in EIAs to ensure that other issues do not erode their importance.

- Certain other countries including the US and Canada (both countries with strong EIA traditions) have adopted the explicit treatment of biodiversity in EIAs and have felt that this issue is of sufficient importance to warrant preparation of specific guidance.

- Explicit treatment of biodiversity in EIAs should help counter (currently often-correct) challenges by non-governmental organisations that biodiversity issues have not been dealt with sufficiently.

- Biodiversity is not limited to the rare habitats and species but has a much wider focus (the variety of all life). Hence for EISs to deal with biodiversity issues they should look not only at impacts on rare habitats and species (widespread current practice) but must also consider wider countryside issues such as impacts on non-protected areas and non-protected species.

- The issue of scale is closely related to issues of the wider countryside. Current EISs tend to have a fine scale i.e. they concentrate on the immediate route of the proposed road and pay very little attention to the context of this route in its wider environment – the ecosystem context of the proposed route and the biodiversity in this region.

## CONCLUSION

The explicit treatment of biodiversity in EIAs generally is a new area in EIA practice with guidance at a relatively preliminary stage (absent in the UK). Lack of guidance on treatment of biodiversity in road EIAs in the UK context is likely to be partly responsible for the failure of EISs to address these issues as developers/ecological consultants have clearly responded to the 1993 guidance resulting in some improvements in the treatment of ecological considerations. Hence, guidance which has been developed and tested would be extremely valuable. For ease of use ideally the guidance could suggest a limited range of core indicators of biodiversity to be assessed within the EIA process. This would also address the cost issue associated with very detailed and extensive survey work, another reason why current EISs fail to address biodiversity adequately. Additionally, if EIS review criteria specifically mentioned biodiversity this would act as a driver to improve current EIA practice in this area.

Development of a method for the systematic treatment of biodiversity issues in the road scheme EIA process and dissemination of information about this method and how to apply it in practice would improve current practice in this area. With these aims, this project developed and tested a systematic approach for the treatment of biodiversity issues and guidance on its use has been produced (RSPB, 2000; Byron, 2000) and is being disseminated.

## REFERENCES

**Byron HJ, Treweek JR, Sheate WR, Thompson S. 2000.** Road developments in the UK: An analysis of ecological assessment in environmental impact statements produced between 1993 and 1997. *Journal of Environmental Planning and Management.* **43(1):** 71-79.

**Byron H. 2000.** *Biodiversity Impact – Biodiversity and Environmental Impact Assessment: A Good Practice Guide for Road Schemes.* Sandy, Bedfordshire: RSPB, WWF-UK, English Nature, Wildlife Trusts.

**Canter, 1997.** *Environmental Impact Assessment (2nd edition),* McGraw-Hill International Editions.

**CAG Consultants and Land Use Consultants, 1997.** *What matters and why? environmental capital: a new approach, a provisional guide,* August 1997.

**CEAA, 1996.** *A guide to biodiversity and environmental assessment,* Canada: Ministry of Supply and Services.

**CEC, 1985.** On the assessment of the effects of certain public and private projects on the environment. *Official Journal* **L175:** 40-48.

**CEC, 1997.** Amended proposal for a Council Directive 97/11/EC amending directive 85/337/EEC on the assessment of the effects of certain public and private projects on the environment. *Official Journal* **L173:** 5.

**CEQ, 1993.** *Incorporating biodiversity considerations into environmental impact analysis under the National Environmental Policy Act,* CEQ.

**CEQ webpage.** The Ecosystem Approach http://tis-nt.eh.doe.gov/oepa/articles /ecosystem.html .

**Countryside Council for Wales. 1997.** *Action for Wildlife: Biodiversity Action Plans - The Challenge in Wales.* Bangor: Countryside Council for Wales.

**Department of Environment, 1995.** *Preparation of environmental statements for planning projects that require environmental assessments: a good practice guide,* London: HMSO.

**Department of Environment, 1983.** *The Manual of Environmental Assessment.* London: HMSO.

**Department of Transport, 1993.** *Design Manual for Roads and Bridges Vol 11: Environmental Assessment.*

**DeLong Jr DC, 1996.** *Defining biodiversity,* Wildlife Society Bulletin **24 (4):** 738-749.

**English Nature 1998a.** *UK Biodiversity Group: Tranche 2 Action plans Vol I - Vertebrates and Vascular Plants.* Peterborough: English Nature.

**English Nature 1998b.** *UK Biodiversity Group: Tranche 2 Action plans Vol II - Terrestrial and Freshwater habitats.* Peterborough: English Nature.

**English Nature 1999a.** *UK Biodiversity Group: Tranche 2 Action plans Vol III - Plants and Fungi.* Peterborough: English Nature.

**English Nature 1999b.** *UK Biodiversity Group: Tranche 2 Action plans Vol IV - Invertebrates.* Peterborough: English Nature.

**English Nature 1999c.** *UK Biodiversity Group: Tranche 2 Action plans Vol V.* Peterborough: English Nature.

**English Nature 1999d.** *UK Biodiversity Group: Tranche 2 Action plans Vol VI.* Peterborough: English Nature.

**HM Government, 1994.** *Biodiversity: the UK Action Plan.* HMSO.

**HM Government, 1995.** *Biodiversity: the UK Steering Group Report Vol I: Meeting the Rio Challenge.* London: HMSO.

**HM Government, 1995.** *Biodiversity: the UK Steering Group Report Vol 2: Action Plans.* London: HMSO.

**HM Government, 1996.** *Government response to the UK Steering Group Report on Biodiversity* London: HMSO.

**IUCN (The World Conservation Union) webpage.** *http://iucn.org/themes /economics.*

**London Wildlife Trust. 1998.** *Biodiversity Action Plans: Getting Involved at a Local Level.* London: London Wildlife Trust.

**Munn RE, 1979.** *Environmental impact assessment. New York:* John Wiley.

**RSPB,WWF-UK, English Nature, Wildlife Trusts. 2000.** *Biodivesity impact – Biodiversity and Environment Impact Assessment: A New Approach.* Sandy, Bedfordshire: RSPB, WWF-UK, English Nature, Wildlife Trusts.

**Sadler B, 1996.** *International study of the effectiveness of environmental assessment final report - environmental assessment in a changing world: evaluating practice to improve performance,* CEAC and IAIA.

**Scottish Biodiversity Group. 1997.** *Local Biodiversity Action Plans: A Manual.* Edinburgh: Scottish Office.

**Takacs D, 1997.** *The idea of biodiversity: philosophies of paradise.* Baltimore and London: The John Hopkins University Press.

**Treweek JR, Thompson S, Veitch N, Japp C, 1993.** Ecological Assessment of Proposed Road Developments: A Review of Environmental Statements. *Journal of Environmental Planning and Management.* **36:** 3.

**UK Local Issues Advisory Group, 1997.** *Guidance for Local Biodiversity Action Plans: Notes 1,2,3 and 4.*

**United Nations Conference on the Environment, 1992.** *The UN Convention on Biological Diversity.* IUCN (The World Conservation Union) A guide to the Convention on Biological Diversity.

**United Nations Economic Commission for Europe, 1991.** *Policies and systems of Environmental Impact Assessment.* New York: UN (ECE/ENVWA/15).

**Wilson EO ed, 1988.** *Biodiversity.* Washington: National Academy Press.

**Wynne G *et al.*, 1995.** *Biodiversity Challenge: an agenda for conservation in the UK (second edition).*

# 7

# Transport and sustainable development

## C FINDLAY

As an engineer my day to day work is more and more tied into the processes and challenges of developing a sustainable transport solution which minimises damage to the environment and in many instances brings improvements.

Much of the strategic road network in the UK was constructed to facilitate the art of war. The Romans have clearly left their mark, but so of course did others, including General Wade north of the border to quell the Scottish Rebellion. The World Wars also necessitated the co-ordination and maintenance of excellent communications between our manufacturing centres and the ports.

At that time construction and maintenance of this country's strategic road network took little or no account of environmental issues. In design and construction, the Engineer was all-powerful. His skills related to horizontal and vertical alignment, and strength of materials etc. For example at local Authority level, the 1819 Report of The Select Committee of the House of Commons on the *Science of Road Making* stated, *inter alia*: *"...each County should appoint an executive officer as a County Surveyor whose services would be amply remunerated as only persons of superior ability were to be considered"*.

A further recommendation was that there should be efficient training of engineers in the science of road making. (But note the emphasis on surveyor). In the post war years it has to be said that road construction continued with little if any thought being given to the environmental consequences. One of the favoured textbooks of the title was *Road Making and Road Using,* by Salkield & Pressey, first published in 1947 and significantly the only reference to the environment is in the section dealing with the noise implications of various surface treatments.

Through the 1950s, 1960s and in part the 1970s the various engineering publications on the science of road construction and maintenance continued to centre on design features and the scale of the construction necessary to accommodate the expected traffic growth.

However by the late 1970s and early 1980s, traffic levels had grown to such alarmingly high volumes, and congestion in our cities was beginning to reach such unacceptable levels, that the concept of 'predict and provide' i.e. predict the traffic growth and provide the necessary infrastructure to accommodate it, began to be questioned by many professionals. Questions asked included:

• Where would such a policy take us?

• What was the economic and financial case for the continuation of such a policy?

• What were the implications for our environment?

However at that time the Government's road traffic forecasts dictated policy. The policy could not be challenged, even at Public Inquiry, and hence some difficult, but in hindsight some very helpful, confrontations occurred during a number of road inquiries. A number of individuals spent a considerable amount of their time being ejected from many public inquiries dealing with new road schemes. Their basic thesis was that building more roads to accommodate yet more traffic without taking full account of the environmental implications was not a sound basis on which to plan their nation's transport infrastructure. Their efforts, linked to the emerging belief that UK transport planning was indeed heading in the wrong direction, resulted in a major change in policy through the publication by the Government of the *1989 Road Traffic Forecasts*. This Government publication was a watershed in transportation policy development in the UK. The traffic forecasts themselves still gave great cause for concern with traffic levels by 2015 expected to be in excess of 200% greater than levels at that time. Very significantly, the Government indicated that such forecasts were only their 'best estimates' of what they believed would occur: the forecasts themselves were not policy and consequently it was the responsibility of each traffic authority i.e. the Department of Transport, County Councils and Metropolitan Areas to make and justify their own forecasts.

As a result there were a number of very important publications including various reports by the Standing Advisory Committee on Trunk Road Assessment (SACTRA) together with the creation of transport plans at county level, which pointed a clear direction away from 'predict and provide' philosophy; earlier assumptions of providing for unrestrained growth were replaced with a developing consensus in favour of management of demand for travel and a fully integrated approach to travel across all modes. It is of great significance that the recent publication on transport policy by the Institute of Highways and Transportation *Transport in the Urban Environment* has as many sections on the environmental impact of transport as it has on other transport issues.

Despite this significant change in policy and the acknowledgement of the critical importance of the principles of sustainable development, the Highways Agency was still progressing with an ambitious road building programme through the early to mid 1990s. Of particular note were the proposed M25 link roads.

The Highways Agency's argument was along the following lines: the M25 to the west of London is carrying up to 200,000 vehicles/day which is well in excess of its design capacity and given the strategic importance of the M25 it is necessary to add additional highway capacity and in particular between junction 12 (M3) and junction 15 (M4). They postulated that 'Link Roads' were the correct solution to the problem. Surrey County Council, as well as many other organisations, was of the firm belief that such a scheme was totally wrong and for the following reasons:

• The impact on the environment, both in terms of direct effects on local residents and the damage to the local environment was not acceptable.

• The concept was not in accordance with the principles of sustainable development.

• The 'link roads' would not solve the problem.

With respect to the environmental implications, the proposal to widen from 8 Lanes to 14 lanes would have had a very serious impact on the Staines Moor Area of Outstanding Natural Beauty and would have resulted in a serious deterioration in air quality along the M25 corridor, which of course is adjacent to urban areas and a number of schools.

Operationally Surrey County Council also argued that the provision of such additional capacity, and in such a localised manner, made no sense in the context of the limited capacity of the wider highway network. The County Council demonstrated that one of the main outcomes of the construction of the 'link roads' would be the serious overloading of the local road network with all the attendant problems of severance, poor air quality and noise. A clear indication that the tide was really timing against unconstrained highway construction was when in 1996, the Government announced that it would not be taking forward the 'link roads' proposals but would be examining the case for more localised on line widening. A major battle has been won. The 'predict and provide' philosophy has been well and truly buried, with the principal of sustainable transport provision coming to the forefront. It is an exciting time for transportation planning in the UK.

## THE CURRENT AGENDA

With the arrival of a new Government in May 1997, a moratorium was placed on the national road construction programme in the UK. The Government also produced its consultation document *Towards an Integrated Transport Strategy* and a Transport White Paper. The transport debate centred on the following issues:

- At a National level we need a National Transport Plan with clearly defines objectives and which will set targets in relation to traffic growth, use of non-car means of transport, air quality improvement etc.

- At a local level, statutory plans which will respond to the national objectives and targets.

- Land use planning and transport provision co-ordinated in order to reduce the need to travel.

- Consideration to the quality of life which we are striving to create Ensure full integration of all forms of transport.

It is my belief that the processes and the objectives in the Transport White Paper will be satisfied.

We have come a long way – from roads for wars, roads to alleviate traffic congestion, and 'predict and provide' – to where we are now. There is little doubt that the age of large road construction programmes is behind us. The agenda is about sustainability and the mechanisms we will be utilising to achieve that end will now include:

- Traffic control centres.

- Demand management.

- Traffic calming.

- Passenger transport improvements.

- Safer routes for cyclists and pedestrians.

- Parking restraint.

At a national level this will mean more emphasis on measuring safety and accessibility on the trunk road network, and in this context we can expect to see an expansion of the M25 speed experiment, more signalised control of Junctions and much information to travellers before they join the network and whilst they are using it. At a local level, the emphasis will be on promoting demand management and enhancement to passenger transport, through measures such as quality bus partnerships, park-n-ride schemes, dial-a-ride schemes and safer routes for all travellers bicycle and foot. We will also be seeing the 'greening' of our existing roads. Many of our road were constructed in a manner which had little respect for the needs sensitivity of the environment.

I hope the positive changes that have occurred in the UK over the last twenty years towards creating sustainable transport policies and implementing sustainable transport schemes. In tandem with this has been the development of sophisticated environmental evaluation techniques, which now form an integral part of the overall evaluation process. I would now like to set out the principles of the environmental evaluation process as used for new highway schemes and use the recently completed Blackwater Valley Route in Surrey as an example of how good practice can be used to minimise the impact of such schemes on the environment and indeed in bringing 'improvement to the environment'.

## ENVIRONMENTAL ASSESSMENTS

In 1969 the first form of Environmental Impact Assessment (EIA) was established in the USA with the aim of controlling future development and conserving the environment. Since then much of the world has followed suit and in 1976 the UK Government commissioned studies into EIAs. In 1985 the UK Government published its *Manual of Environmental Appraisal* (MEA) which set out a procedure for evaluating the environmental implications of new roads. In 1992 SACTRA was invited to review procedures for assessing the environmental impact of road schemes and in response to their findings the Government published the *Design Manual for Roads and Bridges (DMRB) Vol 11,* which specifically deals with the environmental impacts of highways. It formalises assessment procedures in the key development stage of a new scheme – before programming entry, at public consultation and via the publication of an Environmental Statement. This was a major policy change as prior to this, work concentrated purely on mitigation measures. Now environmental implications are given an equal weighting in the scheme along with the financial, economic and operational issues. The DMBR procedures are still in use for national and local schemes.

Within most Counties and certainly in Surrey, major developments are now accompanied by an Environmental Statement. These assessments allow the importance of the environmental effects and the scope of the mitigation measures to be considered by the relevant decision-making authority before development consent is given. Given Surrey's already dense development and the extent of specially designated areas such as Areas of Outstanding Natural Beauty the cumulative impact of even modest developments could have a significant effect on the local environment. Such processes and procedures were followed in assessing the best option, construction materials and environmental instigation measures associated with the construction of the Blackwater Valley Route in Surrey.

## THE BLACKWATER VALLEY ROUTE

The Blackwater Valley Route is an example of how environmental issues have become increasingly important aspects in the development of new road schemes. The route connects the A30/M3 to the A31 across the Hampshire/Surrey boundary. The centre section, a £50 million 4km dual carriageway, was built as a partnership between Hampshire and Surrey County Council, and has been an environmental and engineering challenge. The Blackwater Valley is an area of environmental and recreational importance with many gravel workings, which are now valuable wildlife habitats and fisheries. The Basingstoke Canal crosses the valley on an embankment and is both a Conservation Area and Site of Special Scientific Interest. The Basingstoke Canal was closed for nine months so that a 200m gap could be cut to enable construction of the new aqueduct.

The new road has cut through waterways and wetlands with a rich variety of wildlife, sometimes below the water table. The carriageways cut through massive concrete troughs of up to 2m thick reinforced concrete to prevent flotation of the structures. Excavation for the road ran into a many difficulties; waste mid soil pits, some with medical or military waste, have been uncovered and dealt with, despite the unhelpful complication of ever-present water. It was inevitable that some of the habitats for flora and fauna of national and even international importance would be disrupted.

Initially an environmental assessment was undertaken for the scheme but further more detailed work was carried out following the Environmental Statement stage as awareness of the environmental issues increased during the detailed design stage of the scheme. Although the scheme could have resulted in a potentially negative effect on the Blackwater Valley, extensive and innovative mitigation works costing in excess of £1.5m have minimised the effects and in some areas enhanced the existing environment. Mitigation works were implemented often extending well beyond the immediate confines of the road. Human population densities are high in this area, with local people valuing environmental facilities and landowners (including the British Army on its home ground) taking a keen interest in any changes. The environmentally responsible policies adopted by the County Councils enjoyed extensive local support.

Many aspects of the scheme have been designed such that nature conservation and ecology interests are enhanced. The new 500 metre permanent diversion of the River Blackwater channel has been constructed in line with recommendations from environmental consultants and the Environment Agency, to retain and where possible enhance the ecological value of the river. The features of the new river channel include:-

- A 50 metre corridor, where possible, on either side of the river.

- A meandering effect with small bends and the sculpturing of banks creating pools and riffles.

- Sculpting of the new channel to ensure that it is self-cleaning.

- Setting flood berms at different levels to create different habitats. Some berms are set immediately above normal flow to create marshy conditions, others are set at a level which will occasionally flood.

- Riverbed material from the old channel has been re-used to recreate the original habitats.

- Off channel fish spawning areas have been created.

- Use of hazel hurdles and spilling to prevent bank erosion at key points and encourage flora and fauna to colonise.

## Advance works to protect wildlife and habitats

Once a major engineering operation is underway there is little chance of protecting fauna and flora lying in the direct route of heavy traffic. Consequently great attention was given to reducing the unnecessary loss of wildlife and important plant species and plant groups before construction.

Specialised works started in March 1993, fifteen months before the start of
the main construction with the aim of:

- Creating replacement habitats.

- Rescuing and relocating rare plant species growing on the direct line of
  the main construction site.

- Ensuring important water areas adjoining the route of the road were
  effectively sealed off from the main construction area and protected.

- Enhancing the diversity and value of adjoining water area for wildlife
  and recreation.

- Deterring nesting birds and bats by clearing those trees most likely to be
  occupied in advance of construction works and before the nesting
  season.

- Rescuing fish from threatened lakes in close liaison with the EA.

- Moving snakes, toads, lizards etc. to appropriate alternative areas (many
  created specifically for this purpose). Return journeys to the
  construction site were deterred by the erection of a low-level physical
  barrier.

- A swan rescue.

- A badger survey was undertaken to ensure the area was not occupied.

- Bat surveys.

In Lakeside Park, extensive scrub clearance was carried out to enhance
areas of heathland and flower rich grassland. Overgrown and silted up ponds
were cleared and re-dredged to provide habitats where wetland plants and
turfs from the line of the new road could be relocated. Small areas of
woodland were conserved.

## Ecology and nature conservation

Before and during construction every effort was made to minimise the
harmful impacts and make positive and sustainable changes. From the start
the County Council regarded the protection and enhancement of wildlife
habitats and recreation resources to be central to the scheme.   Where
necessary, this has included the replacement of recreation resources and
restoration of new wildlife habitats.   The commitment to ensuring the
ecology and nature conservation interest of the area was demonstrated by the

setting up a Blackwater Valley Landscape/Ecology Working Group in 1992 and the appointment of a clerk of works with particular expertise in landscape and ecological matters.

## The bats and the cave

One of the greatest innovations of the project has been the building of a bat cave. The Basingstoke Canal, Blackwater River and local lakes provide ideal feeding areas for bats. When it became clear that the new road would interfere with roosting areas, Frank Greenaway, a bat expert was consulted. A man-made bat cave was built, containing a water level hibernating cave linked vertically to a breeding chamber with its own black concrete solar panel. Although too early to assess the bat cave's success bats have visited it and a pipistrelle bat *Pipistrellus pipistrellus,* has been found roosting. The Bat Conservation Trust will monitor its use.

The cave is an artificial soil hill held in place by 'hydroseed' mesh matting. The design of the cave is such that on one side a cliff face has been constructed from sandbags to attract sand martins *Riparia ripiria,* on another, tree root plates saved from site clearance work provide a nesting place for kingfishers *Alcedo atthis.*

## Other measures

Other mitigation measures to enhance the nature conservation interest of the area have included the construction of kingfisher tunnels, bumble bee *Bombus* Spp. chambers and the fitting bird and bat boxes to trees all around Spring Lakes. Even insects have not been forgotten. Dead wood provides a valuable habitat for invertebrates, which in turn help sustain the diversity of wildlife in the area. Rather than burn tree trunks, which is the normal practice on a construction site, stockpiles of dead wood have been left on land adjoining the road to create dead wood habitat.

## Amenity and recreation

Whilst constructing the road the County Council has been committed to protecting and significantly enhancing the formal and informal recreation facilities. Major recreation projects and facilities have included the construction of the final missing link in the Blackwater Valley footpath between Aldershot Road and the Basingstoke Canal. The footpath follows the Blackwater River and is accessible to wheelchair users and pushchairs.

The creation of new areas of public open space and improvements to the existing community park at Lakeside Park, and the internationally important fishing lakes have been protected and enhanced in co-operation with the landowner Lake depths have been reduced and additional shorelines created to improve their general wildlife.

Surrey has ensured that angling facilities have not suffered from construction activities. The County Council and fishing centre owner recognised that the environment and its occupants would be best served by co-operation. The Council has worked with the landowner to protect the internationally respected and privately owned Gold Valley complex of five fishing lakes. There was a need to construct causeways and define a line to prevent contractors causing subsequent disturbance. In return the landowner has provided secure locations for several habitats such as the bat cave.

Recreation has formed a major part of the overall Blackwater Valley Route with the public being encouraged to take part in activities such as walking, cycling and playing and undertake hobbies such as fishing and photography. As the area matures it will become highly beneficial for environmental education within the National Curriculum.

## Landscape

Consultants were commissioned to provide supporting information for the setting up of long term landscape management plans for highway schemes. A range of environmental measures has been undertaken to integrate the schemes into the local landscape. The extent of these measures and their sustainability has highlighted the need for a properly programmed and resourced landscape management strategy.

The Blackwater Valley Route has brought significant environmental change to the narrow open valley, which separates the urban areas of Aldershot and Farnborough in the west from Mytchett, Frimley Green, Ash and Ash Vale in the east. The landscape proposals for the BVR have been seen as a vital and integral part of the overall scheme and as a means of to assimilate the new road into the valley landscape. To help to achieve this, works have been and are being carried out in significant areas beyond the immediate line of the road. Landscaping works have included:-

- Protecting as many trees as possible with fencing during the construction; these mature trees help improve the character of the road.

- Managing areas of existing woodlands, groups of trees and shrubs.

- Advanced planting at Spring Lakes and Willow Park, including the creation of community open space are at Willow Park.

The main landscape contract was carried out during the winter 1996/97 and comprised a total of some 79,000 trees and shrubs and some two hectares of wild flower seeding with a total cost of £650,000. Long term commitment to landscape and conservation works will continue.

## Noise impact

The scheme is adjacent to areas of residential development and areas used for recreation. It was therefore important that the negative effects associated with the noise from road traffic were mitigated. In terms of design the road profile was kept as low as possible which had both considerable noise and visual benefits. In addition the following features were built into the scheme design:

- Noise absorbent blocks were used to line the concrete troughs.

- Porous asphalt has been used throughout, which significantly reduces road traffic noise and increases safety through spray reduction.

- Where appropriate, two metre high earth mounds provide landscaped noise barriers.

The noise mitigation measures have meant that the predicted road traffic noise will be well below the levels for which residential property insulation would be available under the Noise Insulation Regulations.

## The way forward – partnership approach

The Blackwater Valley Environment Working Group was established in 1992. This group has ensured close co-operation with all relevant parties and made sure that landscape conservation measures were implemented in the most appropriate way, based on the maximum possible information and available expertise. Since May 1992 the group has co-ordinated work in the following areas:

- An advanced landscape and conservation programme.

- Habitat protection, management, creation and enhancement.

- Flora and fauna rescue, relocation and protection landscaping.

- Recreation provision – especially the Blackwater Valley Footpath and fishery management.

The group continues to review all existing survey information and commission additional survey work where necessary. The working group will continue to draw on a wide range of expertise and involve a number of interested parties.

## The importance of monitoring

In advance of construction works for the Blackwater Valley Road a full botanical survey of wetlands and other habitats was carried out. New ponds were created, existing habitats brought into management and efforts made to conserve rare plants and plant communities. Since March 1993 there has been annual monitoring of these habitats to provide guidance for further work and management. Monitoring has shown that work in certain areas are proving highly successful with increased variety and numbers of scarce plant species including orchids, water violet and grass vetchling. Aquatic areas have been well colonised and support strong populations of dragonflies *Odenata* ssp. Almost twenty species of butterfly been recorded on the grasslands. The first year results were encouraging, though a number of potential problems were also identified.

The relocation of scarce plants in general is having mixed success. The rarest plant, the gingerbread sedge (first found here in 1888) is flourishing. Some other plants have not colonised. The County Council remains committed to continuing monitoring and management of all improved habitat areas in co-operation with local councils, landowners and the Blackwater Valley Recreation and Countryside Management Service. Monitoring has taken place each year, during July or August at selected sites.

## THE NEXT 20 YEARS

It is clear that the major programme of road construction that we experienced in the late 1970s and through the 1980s will not return. Indeed, it is difficult to imagine any significant level of highway construction whatsoever in the near future other than certain 'gaps' in the nations trunk road network, for example the A3 Tunnel at Hindhead. Where road construction is planned there is little doubt in my mind that the environmental issues will remain to the forefront of the evaluation criteria and when a preferred route is decided on, one will expect mitigation measures of the very highest quality.

Although any new construction programme will be minimal, the highway maintenance programme continues at a very significant level, albeit that the level of investment is not sufficient to maintain the overall standard. Herein lie many opportunities to improve the environment. Measures such as increasing the quality of *in-situ* and off-site recycling of materials must be pursued together with the 'greening' of roads as discussed earlier.

The challenge is real and the processes are understood. What is required is a national drive to continue to reduce the environmental implications associated with all forms of transport.

# 8

# Transport and the biodiversity action plan

## G WYNNE

The RSPB has argued for a suite of changes in policy – to the ways in which transport is planned, infrastructure is built and vehicles are used – to protect biodiversity, as published in *Braking point: the RSPB's policy on transport and biodiversity,1995*. The principle areas that should be addressed include:

- Demand management.

- Encouraging greener transport modes.

- Ensuring better integration between transport modes.

- Encouraging cleaner and more efficient vehicles.

- Improved site protection.

- Changes to instruments (economic and regulatory).

- Changes to appraisal systems and institutions.

Integrated transport policy requires improved and broader policy objectives, and emphasis on delivering these objectives. In addition, there should be improvements in cross-departmental co-ordination and integration.

The Royal Society for the Protection of Birds (RSPB) believes that a strong strategic tier is needed in the decision-making process, while actual delivery for biodiversity takes place at a local level. This is necessary so that the national, UK and international environmental implications of policies and decisions can be recognised and an overview of biodiversity beyond administrative boundaries maintained.

Transport is of particular concern to us for two main reasons. The first is that the impacts of the transport sector on biodiversity are very significant. The second is that the Government is still responsible for very large infrastructure projects that can cause serious damage.

In broad terms, the RSPB welcomes the increasing signals that public funds should be aimed at public transport rather than the roads programme, and road maintenance rather than new build. We would also draw attention to the particular biodiversity impacts associated with port developments, and our work in this area.

## OBJECTIVES FOR TRANSPORT POLICY

Objectives for the transport sector should cover environmental, social and economic issues. The RSPB has long argued for the Government to set environmental objectives and targets for transport which take account of biodiversity conservation. Set out below are the objectives that the RSPB and other biodiversity groups believe are necessary to improve biodiversity conservation in the UK. While overall environmental objectives are set out in general terms, more detailed comments are given on biodiversity objectives.

The RSPB notes that, in addition to environmental and biodiversity objectives, a number of other themes are needed in the transport debate. These themes are widely agreed across core environmental groups working on transport issues. These should include:

- Better integration between policies, for example most immediately through local transport policy and traffic reduction targets.

- Fairer costs, including changes to fuel taxes, private non-residential parking etc.

- Greater choice, particularly emphasising the role of non-motorised and public transport.

- More inclusion of people who find public transport difficult or threatening to use.

- Reclaiming public spaces for people on foot and bicycles.

- Better transport for safer communities, for example concentrating on the issues of severance and speeding.

**Transport impacts on biodiversity**

The transport sector has a number of impacts on biodiversity, both in terms of new build and through operational impacts, set out in the matrix below:

|  | *Primary* | *Secondary* |
|---|---|---|
| Direct | • Habitat loss to transport infrastructure development<br>• Road kills | • Transport infrastructure inducing more traffic and more development in new areas leading to habitat loss and land use changes |
| Indirect | • Habitat fragmentation<br>• Noise pollution impacts on wildlife, notably songbirds<br>• Disturbance effects of artificial lighting on birds/insects/bats<br>• Water pollution in run-off<br>• Local air pollution effects on plants and animals<br>• Effects of spillages<br>• Effects of litter on animals<br>• Effects of roadside verge management practice | • Contributions from the transport sector to emissions that lead to climate change and acid deposition (NB: acid rain and 'nutrification') |

As a result a range of environmental objectives should be set for the transport sector, including:

- Reduced impacts on biodiversity (see below for more details).
- Improved air quality – note human health issues.
- Reduced noise disturbance.

- Reduced impacts on landscape and amenity.
- Improved safety.
- Reduced congestion.

## BIODIVERSITY OBJECTIVES AND TRANSPORT

While we have set out objectives in some detail below, we have not indicated how they should be implemented. Many, if not all, of them should be fully incorporated into existing systems such as Environmental Assessment. These objectives divide into two groups: reducing the effects of new infrastructure; and reducing the effects of operations.

### New infrastructure

No damage to or destruction of internationally important sites. This includes sites protected under European law including designated and candidate Special Protection Areas under the EC *Wild Birds Directive* and Special Areas of Conservation to be designated under the EC *Habitats and Species Directive* which together make up *NATURA 2000*, a pan-European network of protected areas. Also included are sites designated under International agreements such as Ramsar sites, Biosphere Reserves and World Heritage Sites.

The EC *Habitats and Species Directive* is implemented in Great Britain by the Conservation (Natural Habitats, &c.) Regulations 1994 (the Habitats Regulations). Regulation 6 defines the DETR as a competent authority for the purposes of implementing the Directive.

### New proposals

In taking decisions on proposals which could affect a SPA or SAC, the DETR is obliged to follow the steps laid down in Article 6(2)-6(4) of the *Habitats Directive*. These are implemented by Regulations 48 and 49 of the Habitats Regulations. It is Government policy that these regulations also apply to potential SPAs and candidate SACs.

In putting forward proposals for new infrastructure the DETR must assess whether the project is likely to have a significant effect on a SPA or SAC (either on its own or in combination with other plans or projects). In addition to direct effects it is essential that the DETR considers those effects which may arise from projects located some distance from an SPA or SAC e.g. effects on hydrology, airborne pollution, noise pollution etc. The views of English Nature and the Countryside Council for Wales will be important in helping to assess the significance of such effects.

If it is considered that the project is likely to have a significant effect, it must be subjected to an appropriate assessment. Such an assessment must consider the project's implications for the site's conservation objectives i.e. the reason the site has been designated, in order to assess whether the project will adversely affect the site's integrity. Planning Policy Guidance 9 Nature Conservation defines site integrity as "...*the coherence of* [the site's] *ecological structure and function, across its whole area, which enables it to sustain the habitat, complex of habitats and/or the levels of populations of the species for which it was classified.*"

The RSPB recommends that a full environmental statement should be required for all road schemes in line with the RCEP recommendation in 1994. A range of techniques including EA are currently used, although not always to the highest standard. The key thing is that, overall, these should amount to an appropriate assessment for the purposes of the Directive i.e. predicting the significance of the impacts, mitigation options, alternatives and so on.

The assessment should consider whether there are any conditions or restrictions that can be applied which would mitigate any potentially damaging impacts. If such damaging effects cannot be mitigated then the project should be refused unless the following criteria can be met.

First, DETR must be satisfied that there are no alternative solutions which would have a lesser impact e.g. different route options, traffic management schemes etc.

If no alternative solutions exist, the DETR must satisfy itself that the project must be carried out for imperative reasons of overriding public interest (including those of a social or economic nature). The exception to this is those sites, which support priority habitats or species (as defined in Annex I and Annex II respectively of the *Habitats Directive*). In such cases the only reasons for which a project can over-ride the protection of the European site are those relating to human health, public safety, beneficial consequences of primary importance to the environment or which the European Commission consider are imperative reasons of overriding public interest.

If permission is granted which would adversely affect the integrity of a SPA or SAC, then the Secretary of State must ensure that the necessary compensation is provided to secure the overall coherence of the *NATURA 2000* network.

## A strong presumption against damage or destruction

Nationally important sites for nature conservation include all Sites of Special Scientific Interest (SSSI) notified under the *Wildlife and Countryside Act 1981* and Areas of Special Scientific Interest (ASSI, the equivalent in Northern Ireland).

The general principles controlling development affecting international sites (see 2.2.1 above) should apply equally to national sites. All proposals, which are likely to have a significant effect on an SSSI/ASSI, should be subjected to a full environmental assessment, in line with the RCEP recommendation of 1994. This should apply to those proposals which are some distance away from an SSSI/ASSI but which might have a significant effect upon it. Damaging proposals should not be approved other than in exceptional circumstances i.e. where it can be proven that the proposal is in the overriding public interest and that no alternative solutions exist. It is for the applicant to demonstrate why permission should be granted and, through the assessment, why such alternatives are not practicable. If damaging development is to be permitted, then appropriate mitigation and compensation measures should be required. A presumption against damage to or destruction of habitats or species (including the habitats of said species) listed in the EC *Wild Birds Directive*, EC *Habitats and Species Directive, Biodiversity Action Plan* or *Wildlife and Countryside Act 1981* or *Wildlife (Northern Ireland) Order 1985*.

Protected areas cannot exist in isolation and will not, in themselves, deliver the conservation of biodiversity in line with international commitments and obligations. In order to achieve this, it will also be necessary to protect species protected by European and national legislation in: the wider countryside (i.e. outside *Natura 2000* sites); areas of land adjacent to, or between sites protected by European law which act as corridors or stepping stones (see Article 10 of the *Habitats Directive*) or as a buffer. Such protection should also apply to those habitats and species listed on the short and medium lists in the Government's *Biodiversity Action Plan*.

Accordingly, there should be a presumption against the destruction of, or damage to, such areas and species. Consent should only be granted where it can be demonstrated by the applicant that the proposal will not prejudice the achievement of favourable conservation status of the feature(s) of interest (following the definition given in the EC *Habitats and Species Directive*) at the regional, national and local level A presumption against damage to, or destruction of, other identified sites of wildlife importance. Identified Wildlife Sites help to support the national network of SSSIs and contribute significantly to the conservation of biodiversity.

They are identified, usually by local authorities, in accordance with nationally applicable criteria, with the support of the Wildlife Trusts. The Government should facilitate the provision of advice to local planning authorities supporting the protection of Wildlife Sites. Similarly, the Department of the Environment, Transport and the Regions should not implement any plans or proposals, which would damage or destroy such sites, except in very special circumstances.

## Operations

Development and management of transport operating systems must have regard to the reduction of environmental impact. Targets for the reduction of such impacts must be challenging, but flexible enough to take account of improved scientific information.

### *Targeted reductions of $NO_x$ and other air pollutants such as $SO_x$*
Current air quality targets under the UK National Air Quality Strategy are set primarily to reduce the impacts of air pollution on human health. Issues that affect biodiversity, such as long term nitrogen pollution, are not directly addressed. Transport is responsible for a variety of air pollutants which affect biodiversity – and most notably oxides of nitrogen. Specific targets for the transport sector are needed to reduce air pollution that is having impacts on biodiversity – including $NO_x$, $SO_x$, VOCs, black smoke, ground level ozone, dust and particulates.

### *Targeted reductions of $CO_2$ and other greenhouse gases*
Transport emissions currently account for more than a quarter of the UK's $CO_2$ emissions, and this proportion is increasing. Specific targets to reduce $CO_2$ and other greenhouse gas emissions from the transport sector are needed and appropriate policies must be implemented swiftly.

### *Targeted reductions of other pollutants, especially water pollution*
The current levels of pollution, most urgently from run-off pollutants and from diffuse sources, need to be assessed, targets set to reduce them and measures put in place to meet these targets. Programme of accelerated research on issues such as lighting, noise, fragmentation etc. Issues such as lighting, noise, and habitat fragmentation, need further scientific research in order to design compensation and mitigation packages when new infrastructure is built, and to reduce operational impacts.

## TRANSLATING OBJECTIVES INTO POLICIES

Two key changes are needed to meet the detailed biodiversity, and more general overall environmental objectives set out above. First, the overall level of traffic must be reduced, and second the emissions from each of the vehicles that continue to make up the transport system must also be reduced.

In the long term, this will require a land use planning system that more robustly prevents the sort of development that increases traffic. Welcome moves in this direction are already evident, but more will be needed to prevent out-of-town shopping developments, and in particular, housing developments that actually increase traffic levels.

In the short term, modes other than the private car need to be encouraged, along with measures that increase vehicle efficiency, and measures that reduce the attractiveness of using the private car in the first place.

Encouraging modes of transport such as buses, trains, walking and cycling will require substantial changes to the way in which the institutions that run the transport sector are prioritised. Public transport should be made a much higher quality service by:

- Creating and extending public transport networks (including integration between modes such as bus and rail).
- Improving information provision.
- Speeding up public transport, making it more reliable.
- Making public transport more accessible.
- Targeted fares reductions.
- Improved safety.

The above public transport 'carrots' will need to be accompanied by various 'sticks', including the continued fuel duty escalator, changes to parking policies and potentially the use of congestion charges, particularly in urban centres. Other measures that improve vehicle efficiency are also needed; including changes to the Vehicle Excise Duty system and changes to fuel efficiency regulations and standards that encourage cleaner, more efficient vehicles.

# 9

# The constructor's experience: the designers response

## C LIVINGSTONE

This paper gives a view of the Civil Engineer's role when a project is taken forward that will necessarily affect wildlife. The benefits that we currently enjoy from our society are heavily underpinned by the built environment. Whether it is power stations, telephone networks, sewerage systems, railways or roads. As members of society, we have provided the need and the demand for these works. I believe the Civil Engineer's role is to provide the facilities that society needs with the minimum possible impact on the environment. We hope that this paper will demonstrate that we take our environmental responsibilities seriously and do not lightly place concrete or tarmac in the countryside.

The Government is currently undertaking a wide-ranging review of all modes of transport leading to an integrated approach to transport. This will no doubt lead to a change of emphasis in terms of methods of transportation. We do not however believe that all road development will cease. In our view roads will continue to play a very important role in an integrated system. We also believe that we will continue to struggle with the impact of roads on flora and fauna for some time to come.

Currently a major highway can take up to 10-15 years to complete all its planning and design stages. The main elements of the process are set out below:

- Project identification.
- Route selection – Public consultation.
- Statutory procedures – Public inquiries.
- Detailed design.
- Construction.

We would particularly highlight the two main stages where the public is invited to become involved. These are at the route selection stage, where a small number of suitable routes are shown to the public to gain their views and preferences. The second is the Public Inquiry stage where matters of more detail are discussed, usually at length.

This is the overall framework within which a major highway is developed. It seems very simple and straightforward. It is in practice very complex and time consuming. The process has been developed over many years in response to public opinion and the need to explore fully the impact of any road scheme on the environment.

The major reference for the highway designer on all issues related to ecology and environment is the *Design Manual for Roads and Bridges. Volume 11, Environmental Assessment* (DoT, 1993). This governs the approach and level of detail needed at each stage of the planning process. This manual has been drawn up to ensure projects are prepared in a professional, systematic and consistent manner. It also ensures that UK and European legislation covering environmental assessment are complied with. The most important statement in the manual is in the introduction: *"Environmental Assessment should be conducted by people with sufficiently relevant experience"*. The importance of this statement cannot be over emphasised. Specialist ecologists are now important members of the design team with a great deal to contribute.

Flora and fauna do not follow artificial boundaries and take every opportunity to colonise suitable habitats wherever they are. It is therefore unlikely that a route for a new road can be found that has no impact on significant wildlife. The task for the design team, including the ecologist and highway designer, is to develop a route that has the minimum impact on the environment while providing an acceptable highway. As well as ecology a wide range of other considerations have to be taken into account as a new highway is being planned. These include, archaeology, landscape, visual intrusion, impact upon people, safe road design and cost.

There are often many competing issues that must be resolved. Whether it be those raised by local residents, archaeologists, or ecologists. We must also not loose sight of some of the real reasons why a new road is usually being promoted (e.g. road safety and the removal of heavy traffic from unsuitable roads through populated areas). We are often presented with real dilemmas when we have to balance the impacts on people with those of archaeology or ecology.

The process that we (the constructors) now adopt is one of investigation and consultation. We consult widely with all competent and professional bodies. These usually include: English Nature, The Environment Agency, County Council Ecologists and Regional Wildlife Trusts and the like.

This contact is essential if real problems are to be identified early when appropriate actions can be taken to address them. The overall mechanics of the process are well understood by all concerned and follow what are now a classic methodology: Evaluation – Avoidance – Mitigation.

The evaluation process takes many forms. In the initial stages it is based on desk studies and discussions with statutory bodies and local interest groups. Later in the planning process it extends to major fieldwork.

Clearly the best action to take if an ecologically sensitive area is discovered, is to avoid it if at all possible. This is the obvious first choice. However, this is not always possible within the restraints with which we often have to deal. The second approach is to reduce the impact of the highway on the ecology as much as possible and then to provide further mitigation measures. These decisions are not made in isolation but are usually the subject of extensive discussions with bodies such as English Nature and the Environment Agency.

## Examples from the Newbury Bypass

The first is what has become a symbol of the Newbury Bypass project and that is the Desmoulin's snail *Vertigo moulinsiana*. Prior to the start of the work, English Nature had undertaken a study of the chalk-land streams in this area of Southern England to determine if there were grounds to list parts of them as a Site of Special Scientific Interest (SSSIs). In this process, along with other information, the Desmoulin's snail was found to have several colonies along both the River Lambourn and Kennet. One of these colonies was discovered on the agreed alignment of the Bypass on the River Lambourn. It was not practical to move the alignment to totally avoid the impact at this location as the restraints on route alignment in this area were significant. A strategy was developed with English Nature to mitigate the effects of the works on the snail habitat. Firstly the impact on the colony was reduced as much as possible by minimising the earthworks using a combination of retaining walls and steeper slopes.

Secondly an adjacent area was identified which could be used to create the right conditions for the snail. It was agreed that an area three times larger than that directly affected had to be provided to act as a safety measure as there was not precedent for this kind of work. It was also agreed that a monitoring programme should be put in place to confirm that the measures taken were successful. I am pleased to report that two years after the work was completed surveys show that the snails are breeding well and expanding into their extended habitat. Monitoring will continue for some time to confirm that the early results have been maintained.

**The second example, the Adder Bank**

During the investigations an old bank was discovered that was a major hibernation site for adders *Vipera berus*, in a woodland area at the southern end of the project. It would have been perfectly legal to wait for the adders to emerge in spring and move off to their summer range and then remove the bank. However, the ecologists felt that it provided an important resource for adders in this area. The Highways Agency readily approved the proposal to create a new bank adjacent to the new road. Unfortunately there are no standard highway details for adder hibernation sites, so we designed our own in the field. Hollow concrete blocks were placed on a free draining gravel bed, and the blocks were loosely filled with bark chippings to provide a more natural environment. Following some interesting technical debates on the coefficient of friction between the snake and plastic pipe, plastic was used for entry tubes. Recent surveys show the bank is now being used as a new hibernation site.

During fieldwork ecologists discovered a large black bodied cricket in an area of old pasture. Early identification was that it was the English field cricket *Grillis compestris*. If this had been the case it would have been only the third known site in the UK and would have been of the utmost importance. However, following examination at the Natural History Museum, it was identified as a European exotic *Bimaculitis* spp. This incident highlights the need to research fully and understand any ecological issue before taking any necessary action.

## ACKNOWLEDGEMENTS

The assistance of The Highways Agency, English Nature and the Environment Agency in preparation of this paper is appreciated.

## REFERENCES

**Department of Transport. 1993.** *Design Manual for Roads and Bridges Vol 11: Environmental Assessment.*

# 10

# The constructor's experience: the contractor response

## R JONES

Due to the very nature of our activities as contractors, everything we do has an impact to some extent on the environment. The challenge we face is dealing with the environmental issues associated with constructing Works in a way that lessens that impact as far as possible.

## ENVIRONMENTAL CRITERIA AND MANAGEMENT

How do we manage our business to assure compliance with our responsibility for the Environment, particularly when have a plethora of legislation and requirements to deal with?

Contracts contain stipulations on environmental matters. As far as roads are concerned the Highways Agency and the designer are very proactive in assuring full attention to environmental matters from concept to completion, including the statutory approval stage. Also we, as the contractor, must ensure compliance with all requirements, laws and regulations, whether specifically mentioned in the contract or not.

The first essential is setting up of a framework whereby criteria are identified, understood, complied with and monitored; the objective being to integrate environmental considerations into our business consistent with ISO 14001, the *Specification for Environmental Management Systems*. To this end we operate a system producing customised specific *Site Environmental Management Plans* to promote a culture from top to bottom so as to ensure the message is passed to the work-front, is understood and operated. We put both our senior and operational staff through formal training in the procedures, on all our projects.

## SITE ENVIRONMENTAL MANAGEMENT PLANS

These set out:

- Environmental Aims.

- Site Environmental Policy.
- Environmental Targets.
- Information from site environmental investigation.
- Legislative requirements.
- Environmental Acts of Parliament.
- Environmental Regulations.
- EC Directives.
- Conventions.
- Particular Environmental features incorporated in Contract.

**Environmental Aims**

The aims would commonly include:
- Minimisation / Elimination of pollutants.
- Minimising impact on flora and fauna.
- Effective and regular liaison with appropriate environmental organisations.
- Reduction energy consumption.
- Maximisation of reuse / recycling of materials.
- Consideration of environment in selection of materials / methods.
- Promotion of environmental awareness amongst workforce / neighbours.

**Environmental Targets**

The targets would commonly include:
- No pollution incidents.
- No complaints from local community.
- No unnecessary loss of or damage to wildlife or habitats.
- No unnecessary wastage of resource.
- Achievements of 100% success in ensuring all operatives receive continuing environmental training from induction to work completion.
- Generation of environmentally responsible culture.

To assist and advise management, we have a team of advisers whose role is to promote and develop the Company Policy. During the contract, we continue with the training programme to ensure newcomers are aware and to act as refreshers and conduct formalised monitoring.

From these criteria, an environmental risk assessment – 'Identification of Environmental Issues' – is carried out to define what we can do to lessen the risk, acting as both a monitor and audit instrument.

## Environmental Risk Assessment

Environmental risk assessment is achieved by way of an 'Environmental Impact/Opportunity matrix', where the integrated plan will specify responsibility for achieving objectives and targets – also noting the means and time frame for achievement.

- Systematic examination of every Source of environmental impact or opportunity is carried out and recorded.
  These will comprise the aims and targets previously referred to.

- Risks are rated on a Priority system :
  Probability and Consequence on 1 to 5 scale.

- Product of the two gives Priority, on a scale by which the impact of the many factors can be measured.

- Priorities are reviewed - Concentration and effort applied as to how we can reduce the Probability of a risk, and whether the consequence can be reduced.

- Opportunities are reviewed.

By means of continuous improvement and proactive questioning of operations we use our best endeavours and assure the maintenance of a concentrated attention throughout the whole site structure, to assure responsible environmental management. Having set up a system to ensure compliance with legal and regulatory obligations, our works must be programmed so as to meet those obligations.

## PROGRAMMING IMPLICATIONS

- Start date - Period for Completion.
- Environmental Constraints.

- Weather and Season Dependency.

- Physical Constraints.

Programming must consider all facets of interface difficulties.

- Site clearance must be complete before earthworks can commence.

- Some drainage must be completed before earthworks, other drainage after.

- Earthworks are mainly carried out during fine weather, to assure suitability.

- Earthworks must be completed before the paving operations start.

- Structures must be complete before earthworks and paving can be completed.

- Paving is preferably also carried out in good weather - certainly not when freezing.

All these are subject to the vagaries of wind and weather. These programme constraints must be considered in parallel with the environmental constraints.

## Criteria affecting fauna

A matrix of the primary criteria affecting fauna shows the restrictions relating to any disturbance. First of all, the importance of the Contract start date can be seen:

- If there is tree clearance to do – breeding birds preclude March to July.

- Streambeds should be left during March to June, for amphibians to breed.

- Bats must not be disturbed in the nursery season early summer.

- Badger areas are available only from July to November.

- At all times, the quality of water diverted or pumped must be maintained at the highest level to avoid damage to river life, and to maintain the integrity of the aquifer.

The client and contractor will agree the start date so as to minimise impact on wildlife. The contractor will similarly have to arrange his detailed programme throughout the contract period around the environmental constraints, sympathetically together with all the other physical restraints involved as outlined. Another crucial programming aspect is that of dealing with water. The protection of watercourses and groundwater quality is assured by discharge licences, obtained from the Environment Agency. Anticipated pumped volumes are required, entailing detailed consideration of ground water flow characteristics and weather probabilities – all before work starts. The contractor must assure strictly controlled water quality with limited suspended solids before discharge to watercourses or aquifers. This may well involve pumping water through lagoons and settling tanks, or through stone filter media, involving substantial temporary works and management.

Having set out the programme, what else does the contractor set out to achieve?

## BEST PRACTICE

Whilst we are bound to follow a Specification for the Works, there are certain options which can be implemented to alleviate the effect of roads on the environment, by a choice of working methods in conducting the operations. Primarily in relation both to our duty of care and operation of our own culture of 'Best Practice' we try to go a step further when we can to minimise impact on the environment.

### Temporary works?

The choice of temporary works can be sympathetically dealt with. Extended span Bailey bridging may be used to carry temporary access routes well clear of riverbanks to avoid interference.

### Piling - Precast or bored ?

A lessening of impact can be achieved by choice of the piling method. In vibration sensitive areas, bored piling may be the best option. This method can involve the use of Bentonite to stabilise bores, which may not be desirable next to watercourses due to pollution risk, in which case precast piling may be preferable.

### Imported materials?

Every effort is made to maximise the material usage available within the Site, on the basic precept that every cubic metre of material that can be gained is one less metre to be taken to tip and one less metre to import. We utilise ground stabilisation techniques to reduce the requirement for imported high strength materials.

In addition we process indigenous granular materials from the excavations so as to avoid import, saving impact elsewhere and reducing delivery traffic. The effects of this can be dramatic, for example at Newbury where the import of 500,000 cubic metres was avoided by careful use of site won material. This also proved beneficial to a water meadow in the Kennet valley on which planning permission had been gained for gravel extraction which we were able to avoid using as a result of the processing.

Another facet of avoiding import is in the blending operations we can use to combine peat with subsoil and topsoil so as to rectify a topsoil shortage for use on verges, slopes and landscape areas.

## Disposal of surplus materials?

We avoid offsite disposal by working with the designer where possible to use surplus earthworks to enhance landscaping or noise mounds within the Site. If tipping is inevitable, then this can also be sympathetically dealt with to avoid intrusion.

On the M66 the problem of disposing of an old tip of 240,000 cubic metres of household refuse was solved by placing it within a clay lined impervious waste containment cell. All procedures were agreed with the Environment Agency to assure protection of the environment against the effects of dust, litter and leachate.

Similarly, at Conwy in North Wales the area of tipping of the dredging arising from the immersed tube tunnel was left as a bird habitat with the advice of the RSPB - now maturing into a reserve, as reported in an issue of *Birds* magazine.

## ACKNOWLEDGEMENTS

The Assistance of the Highways Agency, Royal Society for the Protection of Birds, and the Welsh Office Highways Directorate is much appreciated.

# 11

# Public opinion, wildlife and roads

A JUNIPER

According to the latest UK Government figures traffic is expected to grow by at least 38% over the next 20 years (Anon.1997c). Until recently UK government policy has been to meet such increased demand through road building. A powerful driving force behind the policy has been the perceived economic benefits of road building; indeed the 1989 DoT white paper published under a Conservative Government was titled *Roads for Prosperity* (Anon. 1997d). More recently, however, the Government's Standing Advisory Committee delivered the verdict that building new roads generates traffic and does not necessarily have a positive impact on the economy (Anon. 1998c).

In the three annual roads reviews that preceded 1997 General Election, the Conservative Government cancelled 237 individual road schemes, and in so doing reduced the public's financial commitment to road building from about £24 billion to £6 billion. In addition to pressures on public finances, a principal reason for this massive downscaling in road building was the dramatic shift in public opinion against new roads because of their environmental impact. One of the factors operating here was the perceived impact of new roads on wildlife. Images broadcast on television of bulldozers clearing trees, and green fields churned into Somme-like battlefields with security guards pitted against protesters, challenged the public to re-examine the car and the perceived need for more roads.

In July 1997 the new Labour Government reviewed the road building programme and scrapped a further two road schemes – one of which was the Salisbury bypass. However, despite a gradual shift in thinking towards managing demand, many roads are still planned, putting some of our most in-important wildlife sites at risk. While our most important wildlife areas – Sites of Special Scientific Interest (SSSIs), Area of Special Scientific Interest (ASSI in Northern Ireland - have been designated to conserve UK biodiversity, current methods of protecting these important. sites from roads are weak. Even internationally important sites can be damaged by development if the government considers there to be an 'overriding' economic imperative to this end. At the time of writing over 100 SSSIs are currently threatened, and still at risk of damage, by road construction.

In addition to the impacts of new roads on designated sites, there is the continuing and (in some respects) worsening environmental consequences of existing roads on wildlife. Road kills of birds, mammals and amphibians, the effects of noise and light disturbance and the impacts of air pollution in the form of acidification of sensitive habitats and global climate change caused by the enhanced greenhouse effect are all major examples of such impacts.

Public concern about the threats from road building increased from 29% in 1995 to 33% of those interviewed in 1996 and continues to rise (Anon. 1997e). However, despite such changes in attitude, car use and ownership is on the increase.

Public concern about the damaging effects of roads has been clearly demonstrated in a number of recent anti-road campaigns. These campaigns have mainly been fought over sites but have generated, and been part of, wider debates over a number of related issues. These include the psychological and political components of 'car culture', the economics of road building, the extent to which designated sites are protected and the rights of citizens to undertake peaceful protest.

One of the first, in 1992, to receive significant media coverage was the campaign against the M3 extension through Twyford Down in Hampshire. This road scheme involved land take from St Catherine's Hill SSSI (an important grassland habitat) and River Itchen Water meadows – a fen and meadow SSSI.

More recently the Newbury bypass was the scene of one of the largest direct action struggles in the history of the anti-roads movement. Thousands of protesters took part in the campaign, successfully delaying clearance work for many weeks. By December 1996 the battle against the bypass saw over £15 million spent on security (Anon. 1997a).

The A34 Newbury Bypass is being built through four SSSIs, one Special Area of Conservation and the habitat of a species supposedly protected by the *Habitats Directive*, Desmoulin's whorl snail *Vertigo moulinsiana*. Yet such destruction will save just a few minutes of journey time and have little long-term impact on traffic and pollution in the town. All this at a cost of £101 million (*Wild Places!* http://f:foe.co.uk/wildplaces).

In 1996 Friends of the Earth took the case against the destruction of the habitat of the Desmoulin's whorl snail to the high court. The case was lost but the Judge, Mr Justice Sedley, delivered the opinion that: "*... one can appreciate the force of the view that if the protection of the natural* *environment keeps coming second we shall end by destroying our own habitat"*.

Increasingly this view seems to accord with that of the public who, more and more, regard the environmental impact of new road building as unacceptable. However, understanding of the impact of using existing roads is more equivocal with both car ownership and usage on the increase.

## ROADS AND THREATS TO WILDLIFE

### SSSIs

Friends of the Earth have collected data on those road schemes that currently threaten SSSIs in Britain. Information has been gathered on trunk roads and local roads. Trunk roads are owned by the DETR – they include all motorways and many A roads. Local roads are owned and managed by local councils – either the County Council or Metropolitan Borough. The data were obtained from DETR publications such as the DETRs Trunk Road Programme (a list of road schemes which are either under construction, due to start shortly or are in various stages of planning); planning applications and local plans from local authorities, and Head on Collision publications (This is a series of publications by various Wildlife Trusts detailing road threats to SSSIs in their area). Sites have been classified as threatened where there is land take involved, or the road abuts the SSSI.

Over 100 road schemes threaten SSSIs in this way. However, these direct threats only represent the tip of the iceberg. Many more SSSIs will be affected by noise and air pollution, run-off during construction or use, and pressure from further development created by the road. These are otherwise known as indirect threats. One study (reported elsewhere in this volume), carried out in The Netherlands, found that road traffic significantly reduced densities of breeding grassland birds in open areas - up to 1500m from the road (Reijuen, R & Foppen, R. 1996).

### Trunk roads and SSSIs

At the time of writing 51 planned trunk roads threaten 62 SSSIs. Between 1992 and 1998 16 SSSIs were damaged by direct land take by trunk roads. These included Itchen Valley (Winchester Meadows) SSSI and St Catherine's Hill SSSI damaged by the M3 extension. The Second Severn Crossing, completed in 1997, involved land take from Severn Estuary SSSI and Gwent Levels – Magor and Undy. The crossing is part of the M4 Relief Road scheme in South Wales.

## CASE STUDIES

### Birmingham Northern Relief Road

The BNRR was first put forward by the DOT in 1980 to "...*provide as soon as possible the infrastructure to relieve traffic congestion and to provide an alternative route to the busy section of the M6*" (Anon. 1998d).

Even though, when in opposition, the Deputy Prime Minister John Prescott said that "*I do not think the BNRR will do anything to relieve such congestion on the M6*", the scheme was given the go-ahead in July 1997 (Anon. 1997b).

The proposed BNRR threatens two SSSIS. Despite this it was not cut from the roads programme when the Government recently reviewed it. Work is due to start in 1999, and construction will destroy 27 miles of Green Belt. It will cross part of Chasewater Heath SSSI (a loss of 3% of the total area is predicted), causing what the Highway Agency describes as 'severe' damage (*Birmingham Northern Relief Road* http://sun1.bham.acuk/c.m. tarpey/). It will also cross over the River Blythe SSSI, which will involve bridging the river and loss of wetland (Anon. 1998d).

### M4 Relief Road in South Wales

A 22 km relief road is planned to link up with the existing second Severn road bridge. This will cut across four SSSIs on the Gwent Levels – Nash and Goldcliff, Redwick and Llandevenny, St Brides, Whitson. The road will not only intersect the Levels but also open up the area to further development.

The Gwent Levels consist of an area of low-lying flat land alongside coast between, Cardiff and Chepstow. They cover over 8,400ha and are one of the largest areas of reclaimed wet grassland in Wales. The levels support 18 regionally, and six nationally, scarce plant species, 112 nationally scarce invertebrates, with 12 species (including the hairy dragonfly *Brachytron pratense*) so rare as to be in the *Red Data Book*. The area is also an important breeding ground for the declining redshank *Tringa totanus* and lapwing *Vanellus vanellus* (Anon. 1993).

### LOCAL ROADS AND SSSIs

At present there are at least 32 local roads threatening land take from 38 different SSSIs. Since 1992, four SSSIs have been directly affected by local roads and have been recorded 'in loss and damage' data reported by the country agencies. These country agencies: the Countryside Council for Wales, English Nature and Scottish Natural Heritage publish a set of data in their annual reports indicating loss and damage suffered by SSSIs in the previous year.

### A351 Sandford/Northport Bypass

This potentially damaging road scheme was put forward by N.E. Purbeck Council, Dorset, as a "...*solution to relieve congestion on the existing A351*", between the A35 Upton Bypass in the north and the Worget Road Roundabout in the south, and to remove non-essential traffic from residential areas. In the words of the council: "*It will improve traffic flows to the general benefit of the economy*" (Anon.1997e).

The scheme will involve land take from Holton Heath SSSI, Morden Bog. and Hyde Heath SSSI, Wareham Meadows SSSI and Sandford Heath SSSI – all of which make up part of what is known as the Dorset Heathlands. This complex is one of the major lowland heathland areas in Britain and is of national and international importance for its plant and animal communities (Anon. 1996).

### NON-SSSI HABITATS

A great deal of high quality wildlife habitat that does not qualify for SSSI status (county wildlife sites for example) has also been damaged, or is threatened, by proposed road works. In addition to direct loss of woodlands, heath, grasslands, wetlands and other habitat, the overall quality of the countryside is reduced by fragmentation and disturbance.

Where significant wildlife populations are affected by such schemes, various mitigation measures are sometimes employed. Tunnels that permit the 'natural' movement of animals such as badgers *Meles meles*, toads *Bufo* spp. and newts *Triturus* spp. are one example. Whilst these and other mitigation measures are helpful in convincing the public that roads can be environmentally friendly, the wider negative impacts arising for wildlife from pollution, fragmentation and disturbance remain.

The impacts of fragmentation on both designated sites and the wider countryside might have very profound long-term biological impacts. Isolation renders small populations vulnerable to local extinction, limits natural movements over large home ranges, limits effective migration and dispersal and will be a real impediment to natural systems adapting to large scale changes like climate change. Small areas of habitat are less diverse than large ones and the impact of road deaths on vertebrate populations may be significant.

## WILDLIFE KILLS AND DISTURBANCE

Millions of birds and 20-40% of the UK breeding population of amphibians are believed to die each year on roads. About one million mammals die on roads each year. These deaths are mainly of rabbits *Oryctolagus cuniculus* and foxes *Vulpes vulpes* but also include 100,000 hedgehogs *Erinaceus europaeus* and 47,000 badgers *Meles meles* (about one for every badger family in the country). A growth in road traffic journeys will increase this hazard (although there might be an as yet unquantified relationship between increasing road traffic disturbance and the extent to which animals avoid roads, decreasing deaths under some circumstances).

There is also robust evidence that there are non-lethal pressures that diminish the density and breeding effort of some species near to roads. Dutch research for example (Reijuen, R. & Foppen, R. 1996) showed that of 12 bird species investigated, 7 exhibited reduced densities near roads. The same study showed that for some species there was a relationship between traffic levels and distances from which roads would be tolerated (noisier roads have a bigger impact). Another recent survey showed 26 of 43 species at reduced densities. up to 500m from roads carrying 30,000 vehicles or more per day. Reductions in density within 250m of such roads ranged between 20-98%.

## AIR POLLUTION AND CLIMATE CHANGE

Emissions from car exhausts pollute the air and cause environmental damage. $NO_X$ pollution can damage sensitive ecosystems, whilst ozone, formed from photochemical transformation of vehicle exhausts, can damage plants. The emission of carbon dioxide is linked, with increasing confidence, to the enhanced greenhouse effect and global climate change.

Carbon dioxide emissions from vehicles presently account for about 26% of the total arising from UK sources. If present traffic growth forecasts prove correct, this proportion will rise to between 28-34% by 2020 (176-194 million tonnes).

Data presented by the Intergovernmental Panel on Climate Change (IIPCC) suggests that reductions in emissions from fossil-based energy sources will need of the order of 60% will be required by 2050 if dangerous climate change is to be avoided. Clearly, even if the interim UK unilateral target of a 20% reduction by 2010 is to be reached, then a serious impact must be made on the contribution of carbon dioxide from the transport sector. Although not directly linked to impacts on wildlife, climate change is the most profound threat to biodiversity both in the UK and globally.

The prospect of dangerous climate change is only just beginning to penetrate the popular consciousness and is not so far linked in the popular media to the impact of roads and traffic on wildlife.

## FUTURE PROSPECTS

*"...Our general approach is to look at the transport problems which lie behind proposals for road schemes and then seek solutions which are environmentally sustainable"* Dr Gavin Strang, Minister of Transport (Anon. 1997f).

The last five years have seen dramatic shifts in public opinion away from road building and towards (as yet unspecified) alternative approaches to meeting transport needs. The principal drivers of this shift in public opinion have been images of countryside destruction and increasingly convincing arguments that increasing road capacity does not solve transport problems, or necessarily meet transport needs.

Wider transport debates have ensued, notably centred on the question of road traffic reduction and the role of cars in a modem transport system. This wider discussion has additionally been driven by (among other things) concerns over air pollution (and impacts on human health) and the effects of congestion on the economy. The question of climate change increasingly figures as a relevant issue in many a policy-making forum.

The fact that opinion, policy and politics have shifted is beyond dispute; the effect on behaviour is another matter. Juxtaposed against rising public concern about new roads has been a continuing forecast for more cars, car drivers and car use. Clearly the link between changing attitudes on road building and the directly linked question of personal car use have quite different manifestations in the public mind. It seems that countryside destruction, because of road building, is bad but car use, because it is convenient, is good (or at least more challenging because it affects the individual).

In the July 1997 strategic review of road building, two schemes were scrapped. One of these was the Salisbury bypass - which would have impacted on East Harnham Meadows SSSI, Cockey Down SSSI, River Avon (Proposed SAC), and West Harnham Chalk Pit SSSI. It was concluded by the DETR that the environmental dis-benefits of the bypass would outweigh the benefits to through traffic. Other schemes were reviewed further in July 1998, including the Bingley Relief Road (which threatens

Bingley South Bog SSSI) and the A259 Pevensey-Bexhill scheme (which threatens Pevensey Levels SSSI).

Against the backdrop of somewhat contradictory expressions of public opinion, policy is being developed in terms of plans for specific roads. In wider policy several areas are especially significant.

## Integrated Transport Policy

A lack of integration of planning and traffic policies is partly to blame for an increase in car use - over 40% of new housing has been built outside of existing urban areas and their public transport facilities (Anon. 1997). So, the building of new houses on green-field sites or rural brown-field sites also impacts on wildlife and this is clearly linked to transport issues. If the Government anticipates fighting an effective campaign to save the environment and to avoid congestion, then more effective integration between transport policy and development planning must be achieved.

Decades of planning around the motorist have produced an inadequate and underfunded public transport system. There must be greater investment into public transport rather than towards new roads, along with improved cycling and walking networks. The Government is currently consulting on a new Integrated Transport Policy White Paper. Its conclusions are likely to closely reflect understood public opinion, in respect of how best to deliver transport needs in the context of acceptable environmental impacts. Clearly, generating more images of damaged countryside and ruined wildlife is not something the Government will wish to encourage by pursuing the flawed road prioritisation policies of the Conservatives, but neither will it wish to alienate the millions of car owners who voted Labour. Where the lines are drawn in forthcoming policy documents will undoubtedly reflect this political tension.

## Road Traffic Reduction Bill

A National Opinion Poll (NOP) for Friends of the Earth showed that 79% of the public agreed that the Government should set targets for reducing the amount of traffic on the roads. The Road Traffic Reduction Bill – which would require the Government to produce a national plan to cut road traffic from 1990 levels by 5% by the year 2005 and 10% by the year 2010 has been backed by the Government and, if implemented will be an important step forward for transport policy in Britain (Anon. 1998a). There are, however, concerns that despite backing by more than half the members of the House of Commons and, in-principle, Government support, this Bill might be watered down. This is anticipated following the Government's removal of the legal commitment to reduce road traffic levels by 10% by 2010 and instead committing to decreases in projected increases (i.e. to allow continued traffic growth, but to limit the rate of this growth). This again reflects the extent to which the Government presently judges the public unwillingness to accept the more demanding measures to deliver real decreases in traffic levels.

### Eco-taxing

The countries largest car franchiser, 'Lex', Report on Motoring discovered that of the motorists questioned, 76% said that charges for parking at work would prevent them from driving there (Anon. 1998c). Two possible environmental taxes that would target traffic levels are road pricing and car parking charges. Stockholm is one city in which environmental taxes on motorists are implemented. There, motorists are charged up to 36p every time they pass one of the 90 fee stations situated in 10 zones across the city. This Swedish scheme raises more than £90m a year, some of which is used to fund public transport (Whitelegg. 1997).

There is also a range of possible ecological taxes that could be used to encourage alternatives to car use, but in the UK those proposed are limited. At present, the Government is committed to an annual fuel price increase and in the 1998 budget made a first modest commitment to vary vehicle excise duty to favour smaller, less polluting cars and to increase, in a small way, expenditure on public transport. These initial ecological taxes, although helpful, will not render the UKs road transport sector environ-mentally sustainable. This reflects, to a great extent, the failure of the public to yet accept that car use needs to be reduced.

### CONCLUSIONS

The threats to the UKs wildlife through roads and road use are considerable. Habitat loss, fragmentation, air pollution and direct kills of wildlife continue to. seriously deplete the country's wildlife resources.

Because of this, campaign organisations such as Friends of the Earth, Road Alert, Alarm UK and Earth First! have run high profile anti-road campaigns to draw public attention to the damage caused by the road transport sector. These have achieved a very high media profile and have altered public opinion such that government plans for road building have been massively scaled down.

However, public opinion toward car use remains problematic. The public's continuing attachment to personal car transport remains a major blockage to the implementation of policies to tackle air pollution, climate change and the loss of millions of animals and birds annually through direct collisions with cars.

Reducing car use is a priority for sustainable development. To do this, conservation organisations need to communicate more effectively with the public over the ecological consequences of excessive car use. Friends of the Earth are already engaged in work to achieve this new awareness and other organisations should do similar work. If public opinion is shifted such that the damage caused by cars is more widely appreciated, then moves in Government policy to promote regulations and financial incentives to meet traffic reduction targets would be regarded as more feasible.

The first battles have been won. Conservationists must now create strategies that lead to the generation of a transport sector that is truly ecologically sustainable.

## REFERENCES

**Anon. 1993.** M4 Relief Road (M48) Threatens Levels. *Gwent Wildlife Trust Newsletter.*

**Anon. 1996.** Morden Bog SSSI citation sheet. Peterborough: English Nature.

**Anon. 1997a.** Anniversary of Newbury Protest. *Press release* London: Friends of the Earth.

**Anon. 1997b.** Labour Fails Green Test. *Press release* Friends of the Earth.

**Anon. 1997c.** *National Road Traffic Forecasts.* Department of Transport. London: HMSO .

**Anon. 1997d.** *Unlocking the Gridlock: Discussion Paper 4.* London: Friends of the Earth.

**Anon. 1997e.** Sandford and Northport By-pass and A351 improvements. Dorset: N.E. Purbeck District Council.

**Anon. 1997f.** New Strategic Review of Road Building in England. London: DETR.

**Anon. 1998a.** Campaigners claim traffic triumph as government backs road traffic reduction bill. *Press release* London: Friends of the Earth.

**Anon. 1998b.** Motorists Will Not Be Driven From Cars. *The Times.* London: Times Newspaper.

**Anon. 1998c.** *Transport Intensity and Economic Growth: Interim Report.* The Standing Advisory Committee on Trunk Road Assessment Transport Investment London: HMSO.

**Anon. 1998d.** *Non-Technical Summary Volume 1 of the Environmental Statement for BNNR.* DOT and MEL., Birmingham: Department of Transport.

**Reijuen R. Foppen R. 1996.** The Effects of Traffic on the Density of Breeding Birds. *Dutch Agricultural Grasslands. Biological Conservation* **75:** 255-260.

**Taylor B. 1997.** Green in word. In: *British Social Attitudes.* Hampshire: Dartmouth Publishing Company.

**Whitelegg J. 1997.** *A Guide to Achieving Traffic Reduction Targets in England and Wales.* Liverpool: Liverpool John Moores University.

# 12

# The ecological impact of
# air pollution from roads

M R ASHMORE

Road traffic emits a range of substances into the atmosphere, which may have effects on human health, materials and the natural environment. These include lead and other metals, volatile organic compounds (VOCs), carbon monoxide, nitrogen oxides, particulates, sulphur dioxide and carbon dioxide. In terms of evaluating the potential impacts of these pollutants on the natural environment, the key issues are the concentrations of the pollutant at different distances from the road, and whether the species or communities found close to the road are sensitive to the pollutant concerned. It is also important to consider whether there is good empirical field evidence that proximity to roads, and hence to higher air pollutant concentrations, has an ecological impact. Finally, are there appropriate tools available to allow the ecological impacts of air pollution from planned road developments to be assessed?

This paper firstly reviews some of the key generic issues which are important in evaluating the impacts of pollution from road traffic, and to distinguish the different spatial scales over which these impacts may occur. It then considers the impacts of specific pollutants, evaluates the evidence of ecological impacts, identifies priorities for further research, and considers the implications for future road planning.

## KEY ISSUES

When evaluating the impacts of air pollution from roads, it is important to consider a number of key issues:-

- It is important to distinguish the presence of elevated concentrations of a particular substance (contamination) from evidence that these concentrations have an adverse effect (pollution). Thus, many studies have demonstrated that elevated concentrations of lead are found in air, soils, plants and animals close to major roads, but the evidence that these elevated concentrations have ecological impacts is more limited.

- The effects of pollutants may be cumulative over time. Thus, whereas the impact of opening a new road on wildlife mortality may be immediate, that of air pollution may be much more gradual. Continued emissions, for example, of metals may lead to slow increases in roadside concentrations, while emissions of nitrogen oxides may lead to a gradual increase in the nitrogen content of vegetation and soils close to roads. Biological responses to these gradual chemical changes may be non-linear, only becoming apparent when pollutants have accumulated to specific threshold concentrations. This means that, in terms of air pollution, the full ecological impact of the road building programme, which has taken place over the past thirty years, may only appear in the decades to come.

- The effects of combinations of pollutants, such as those emitted from road traffic, may be quite different from those of the individual pollutants, and this may make it difficult to attribute any effects observed to a particular component of the exhaust emissions. Nevertheless, the potential to link ecological effects to a specific pollutant has important policy implications. For example, emissions of particles and carbon monoxide tend to be high at low speeds, whereas those of nitrogen oxides increase rapidly at high speeds. If nitrogen oxides have greater impacts on a particular community, then measures to reduce local speeds might be beneficial, but if carbon monoxide or particles had a greater impact, then this would not be the case.

- There is large interspecific and intraspecific variation in the sensitivity of organisms to air pollution. Thus, the impacts of a particular road scheme may depend critically on the particular communities and organisms found close to it. Evolution of tolerance to both soil metal pollution and atmospheric pollutants has been clearly demonstrated in grasses, and the demonstration that roadside populations have greater pollution tolerance may provide evidence that pollutants are present in ecologically significant concentrations.

- The effects of air pollutants may be substantially modified by other local factors, such as climate, soils and management. It is important that these, and other, interactions are considered when evaluating the potential impacts of a particular road scheme. In the case of roadside communities, there may be additional specific stress factors, which may interact with air pollutants, such as salt accumulation in soils or the effects of gusts of wind produced by passing traffic.

## LOCAL AND REGIONAL IMPACTS OF POLLUTION

In terms of the impacts of air pollution from road traffic on the natural environment, three types of concern may be identified:-

• Impacts on roadside verge communities.

• Impacts on communities close to roads - the distance over which such impacts may occur is unclear, but is likely to be primarily within 1 km of the road, because of the rapid atmospheric dispersion of pollutants.

• Impacts of regional, or global, pollutants to which road traffic contributes. These may include the contribution of carbon dioxide to global warming, the contribution of nitrogen oxides to regional acidification and eutrophication, and the contribution of nitrogen oxides and VOCs to regional photochemical ozone production.

This paper will primarily address the local impacts of air pollutants from roads - the first two categories of impact. However, it is important to recognise that the broader regional ecological impacts of road traffic emissions - the third category of impact - may be of greater national concern than the local impacts of individual roads. Thus, critical levels of ozone, set to prevent effects on sensitive native species, are exceeded over all of southern and central Britain (UK CLAG, 1997; UK PORG, 1997), and there is evidence of shifts in the competitive balance between grassland species, and effects on tree growth and physiology, at ozone levels found in the UK (Ashmore & Ainsworth, 1995; Ashmore & Davison, 1996; UK CLAG, 1997).

Increased deposition of nitrogen has been linked to a number of ecological changes in the UK, including continued failure of *Sphagnum* regeneration in the southern Pennines, a decline in the upland moss *Racomitrium lanuginosum* in England and Wales, decline of *Calluna* in lowland heaths, and damage to snowbed bryophytes in the central Highlands of Scotland (Bell, 1994, INDITE, 1994, Woolgrove & Woodin, 1996), although in each case, it is likely that other factors, such as changes in management, are also involved. Emissions of nitrogen oxides may also contribute to soil and water acidification.

Clearly, when considering the impacts of new and existing roads, the policy implications of the third category, which relate to overall national transport policy, differ significantly from those of the first or second category, for which the choice of specific local routes, or of local traffic management, may influence the impact on particular sites of conservation value. However, it is also important to recognise that the impacts of a local road scheme may depend on the regional background concentration of particular pollutants. For example, Figure 1 shows an evaluation of the area of the country over which critical levels of nitrogen oxides are exceeded - the critical level being defined as the atmospheric concentration above which adverse effects may occur on sensitive species (UK CLAG, 1997). Road traffic is the most important contributor to concentrations of ground-level nitrogen oxides across the UK (UK PORG, 1997). The density of shading in Figure 1 indicates the prevalence of semi-natural vegetation. Where new road schemes, which will tend to increase local concentrations of this pollutant, are proposed in areas where the critical level is already exceeded, the potential for impacts on sensitive types of vegetation exists. However, it should be emphasised that  exceedance of the critical level only indicates that the potential for ecological impacts exists, not that there will be real adverse effects (UK CLAG, 1997).

In evaluating the local impact of air pollutants from road traffic, I will review two separate issues. Firstly, I will briefly consider the impacts of the specific pollutants likely to of greatest ecological concern, and secondly, I will consider the evidence of a small number of specific types of impact, which may not be attributable to specific pollutants. Much of the evidence, which I will draw on, is considered in more detail in an MSc thesis by Hill (1990). The broad principles of air pollutant impacts at an ecological level are considered by Ashmore (1997).

## IMPACTS OF SPECIFIC TRAFFIC POLLUTANTS

The UK National Air Quality Strategy lists eight pollutants, which are of primary concern, based on their impacts on human health. Table 1 summarises these pollutants, comparing briefly their health and ecological impacts. A final category, nitrogen deposition arising from both nitrogen oxides and ammonia emissions, has been included because of its great potential significance for nature conservation. It is apparent from Table 1 that there are considerable differences in the relative importance of the pollutants and the issues of concern - hence, evaluation of road developments and traffic management schemes in both urban and rural areas will need to consider different air pollution aspects if ecological impacts are considered rather than health impacts. Of the pollutants listed in Table 1, $SO_2$ and CO emissions from road traffic are of no ecological significance, while ozone and nitrogen deposition are regional, rather than local, problems. Thus, this section will concentrate on four pollutants - particles, lead and other metals, ethylene and nitrogen oxides.

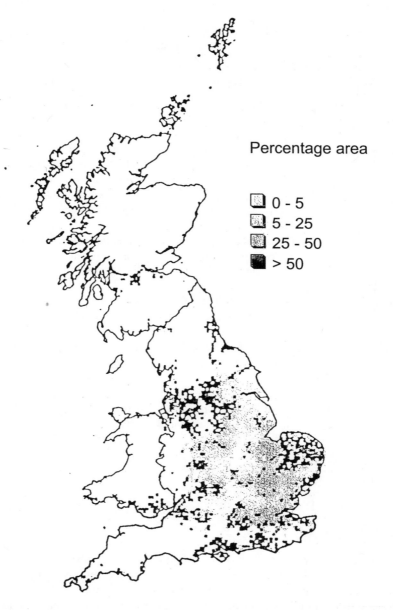

**Figure 1**: Map of the area of the UK where the critical level for nitrogen oxides is exceeded; the white areas are those where it is not exceeded, while the density of shading in the other areas indicates the percentage area covered by semi-natural vegetation. From UK CLAG (1997).

**Table 1**: Comparison of significance of vehicular air pollutants For impacts on health and on ecology

| Pollutant | Health effects | Ecological effects |
|---|---|---|
| Nitrogen oxides | $NO_2$ only of concern | Both $NO_2$ and NO of concern |
| Ozone | Widespread potential effects | Widespread potential effects |
| Particles | Major health concern | Ecological significance depends on chemical characteristics |
| Carbon monoxide | Effects on oxygen-carrying capacity of blood | Little ecological significance at typical concentrations |
| VOCs | Potential carcinogenicity of concern - benzene most significant | Potential effects on flower and leaf drop - ethylene most significant |
| Sulphur dioxide | Emissions from road traffic of limited concern | Emissions from road traffic of limited concern |
| Lead | Effects on childhood intelligence of concern | Effects on plant roots and soil microbial process; increased animal body burdens |
| Nitrogen deposition | Contribution to regional nitrogen deposition not of concern | Contribution to regional nitrogen deposition of considerable ecological concern |

## Particles

Particles are emitted by motor vehicles, and may also be re-suspended from road surfaces. Farmer (1993) provides a detailed review of the effects of dust, including road dust, on vegetation; dust can reduce light interception, interfere with stomatal responses and water relations of vegetation, and cause increases in leaf temperature. Particles can also alter soil or substrate chemistry, change decomposition rates, and affect invertebrates and vertebrates. There is clear evidence of these types of effects in a number of studies of roadside habitats reviewed by Farmer (1993). However, the significance of these effects away from the immediate roadside is unclear, especially since concentrations of the larger particles drop rapidly with distance away from the road. The effects will also depend on the physical and chemical characteristics of the particles.

Trees can be efficient collectors of particles and roadside planting may significantly reduce particle deposition further from the road. For example, a recent study by Freer-Smith *et al.* (1997), in a mixed deciduous woodland adjacent to the M6 in the West Midlands, indicated that the number of particles on the leaf surfaces decreased rapidly as distance from the road increased. A number of studies have shown considerable differences between species in the rate of deposition of lead aerosol to leaves, and this is likely to also apply to other types of particles.

The clearest evidence of effects of road dust on community composition comes from studies of the effects of dust from gravel roads in tundra communities, also reviewed by Farmer (1993), with effects on bryophytes and on acidipholous species being most marked; effects were not found where the road passed over limestone substrates. There is also evidence of effects of point sources of cement dust on local vegetation, but the issue of the effects of dust generated during road construction on sensitive local communities is one that has not been investigated in any depth.

## Lead and other metals

There is clear evidence from many studies that total soil lead concentrations in the immediate vicinity of major roads are significantly elevated due to enhanced aerial deposition. There is a rapid fall in concentration with distance, due to the high deposition rate of lead in the large particle sizes. The lead which is deposited within 50-100m from the road represents only a small fraction of that which is emitted, as the lead in the finer particles has a much longer residence time and can disperse more widely. In some cases, topographical features can result in locally enhanced deposition of the fine particles; for example, Kingston *et al.* (1988) have shown enhanced deposition of lead at woodland edges over 500m from the M25 motorway.

Much of the deposited lead may be tightly bound to the soil matrix, and the extent to which elevated soil lead concentrations are bioavailable, and affect plant and animal communities close to the roadside, is less clear. In the case of vegetation, some of the strongest evidence that these concentrations can have an adverse effect comes from studies which demonstrate that lead tolerant populations of species such as *Plantago lanceolata* (Wu & Antonovics, 1976) and *Festuca rubra* (Atkins *et al.*, 1982) develop close to the roadside.

Development of metal-tolerant populations is good evidence of a significant stress factor, which puts more tolerant ecotypes at a selective advantage. However, lead-tolerant populations are only found close to roads (within 10-20m), and there is little evidence that stress from this pollutant has such effects at greater distances.

In terms of invertebrates, there is good evidence that lead body burdens are higher close to major roads in groups such as earthworms (Ash & Lee, 1980), millipedes and woodlice (Williamson & Evans, 1972), and arachnids (Wade *et al.*, 1980). The evidence that these concentrations have any effects on the survival or population density of these organisms is very limited, although the possibility or more subtle effects on insect behaviour cannot be excluded. Similarly, higher body burdens of lead have been found in small mammal populations captured from roadside sites (e.g. Williamson & Evans, 1972; Jefferies & French, 1972), but there is little evidence that this is associated with adverse effects on their populations. In the case of birds, Grue *et al.* (1986) found higher lead body burdens in starlings nesting within roadside verges, but there was no evidence that these concentrations poses a serious hazard to the bird populations.

The lack of evidence of effects on populations of vertebrates may be due to the insensitive parameters used to detect effects. A recent Italian study (Ieradi *et al.*, 1996) has shown that high body burdens of lead, cadmium and zinc in mice captured in areas of Rome with a high traffic density were associated with an increased frequency of genetic abnormalities. However, it is possible that these effects were linked to high concentrations of mutagenic organic pollutants, which may also be associated with high traffic densities.

## Ethylene

Motor vehicles emit a range of volatile organic compounds (VOCs); while concentrations of many of these compounds are higher in the atmosphere close to roads, the evidence that they have any significant effect on plant and animal life at the concentrations found is very limited. In terms of effects on vegetation, the most important constituent of vehicle exhaust is ethylene (ethene), which is a naturally occurring plant hormone often associated with stress. Controlled experiments using high concentrations of ethylene clearly show that it can cause flower, fruit and leaf abscission, and accelerate leaf senescence. However, whether these responses occur as a result of the concentrations found at the roadside is unclear; one key issue is that plant sensitivity to ethylene is much greater at higher temperatures, but the build-up of high atmospheric concentrations is usually associated with low temperatures. There is also very wide variation in sensitivity between species;

while the most sensitive species, such as certain orchids and marigolds, show damage at concentrations of about 10ppb, many species are unaffected by concentrations which are above 10,000ppb (Taylor *et al.*, 1986).

A study by Sawada *et al.* (1989) attempted to develop a simple model, based on controlled laboratory studies, to predict the effect of ethylene and temperature on leaf shedding by different tree species in urban environments in Japan. The species were grouped into three categories depending on their sensitivity to the pollutant. For those species in the intermediate or tolerant categories, no significant leaf shedding was predicted at any time of year; however, the most sensitive species, which were all evergreen broad-leaved species, were predicted to show significant leaf shedding in the summer months. A similar type of exercise is needed for UK roadside conditions to properly assess the impact of this pollutant; however, it is likely to be broadly true that only a limited number of sensitive species are likely to be affected by concentrations of ethylene close to roads.

**Nitrogen oxides**

Many UK studies have demonstrated the effects of distance from a road on $NO_2$ concentrations, although relatively few have been in remote rural situations. In a recent study, Bell & Ashenden (1997) examined the effect of the distance from different types of road in Snowdonia in winter and summer. Concentrations at a remote site, and at 250 m from the major road examined, were higher in winter than in summer. In contrast, 1m from the main road, concentrations were higher in the summer months, when vehicle numbers are approximately twice those in winter (Figure 2).

Similarly elevated summer $NO_2$ concentrations were found close to a minor road with heavy summer tourist traffic. These data indicate that increased tourist vehicle numbers brought into areas of high nature conservation value may affect air pollution levels close to major roads.

It is important to note that nitric oxide (NO), as well as nitrogen dioxide ($NO_2$,) may be of significance for vegetation (UK CLAG, 1997), although there is little data to allow direct comparison of their relative toxicity at typical ambient concentrations.

However, even if NO is less toxic at the same concentration than $NO_2$, the fact that concentrations of NO close to roads are significantly higher than those of $NO_2$ means that its contribution to ecological effects cannot be ignored. Nitric oxide is the primary form in which nitrogen oxides are emitted from vehicle exhausts, and is then gradually oxidised to form $NO_2$. This means that NO concentrations are often much higher close to roads. For example, Figure 3 shows data from a Swedish study by Sjodin *et al.* (1988) (cited in Hill,

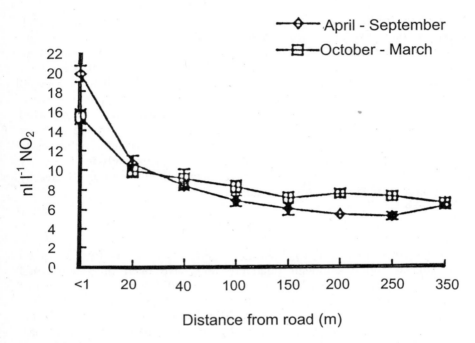

**Figure 2.** Effect of distance from the A5 in Snowdonia National Park on concentrations of NO₂ in summer and in winter. From Bell & Ashenden (1997).

1990), in which the NO:NO₂ ratio was 5:1 in the centre of a rural motorway; NO concentrations fell with distance from the road more rapidly than those of NO₂, but NO concentrations were still twice those of NO₂ at a distance of 80m. These results are consistent with the results of rural monitoring 15m from the M1 in Bedfordshire (Clark *et al.*, 1988), which showed a mean concentration of NO of 116ppb, compared to a mean concentration of 30ppb for NO₂. The data of Flückiger *et al.* (1979) in Switzerland show a very rapid decrease in the concentration of NOₓ (NO & NO₂) from about 500ppb at the edge of the motorway to about 50ppb at a distance of 200m. However, despite this rapid decrease in concentration with distance from the road, the concentration at 200m is still an order of magnitude greater than the NOₓ concentration found at rural background sites in Switzerland.

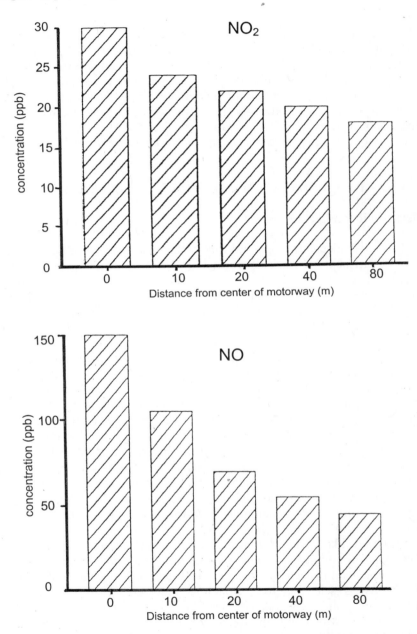

**Figure 3.** Effect of distance from a motorway in Sweden on the concentration of NO$_2$ and NO. From Sjodin *et al.* (1988), reproduced in Hill (1990).

The critical level adopted for nitrogen oxides refers to $NO_x$, rather than $NO_2$, and is set at 15ppb as an annual mean concentration (UK CLAG, 1997). On this basis, the concentrations found at some distance away from major urban roads could still be sufficient to affect sensitive communities. It should be noted high $NO_x$ concentrations may not lead to adverse effects on vegetation directly, but indirectly through changes in pest infestation, fungal infection or sensitivity to climatic stresses (UK CLAG, 1997). Evidence of ecological effects of increased nitrogen deposition close to roads is considered below.

## EVIDENCE OF IMPACTS OF AIR POLLUTION FROM ROADS

### Effects on insect herbivores

There is evidence from several studies that outbreaks of defoliating insects are more frequent on roadside vegetation. For example, Port & Thompson (1980) reported cases of defoliation by lepidopteran species of roadside shrubs, and suggested that the primary reason was the higher concentrations of nitrogen oxides in such locations, which lead to higher leaf nitrogen concentrations (Figure 4) and hence improved food quality for insect herbivores.

Spencer *et al.* (1988) and Spencer & Port (1988) examined the interactions between atmospheric and soil nitrogen and plant and insect performance, using ryegrass with an aphid species. Plants grown in standard soils showed greater biomass and higher nitrogen contents close to the road, and this was associated with improved aphid performance. The effects were particularly marked in soils with low nitrogen status, and there was also evidence that salt treatment enhanced the increase in nitrogen content close to the road, suggesting an interaction between the two stress factors. When plants were grown in soils taken at different distances from the road, those within 1m of the road showed greater ryegrass biomass and higher nitrogen contents than those 6m away, suggesting a further effect though increased soil nitrogen contents close to the road.

The most likely single cause of the increased performance of insect herbivores is the higher concentrations of nitrogen oxides close to the road, since assimilated nitrogen can improve the food quality of the host plant. The work of Swiss scientists (e.g. Braun & Flückiger, 1985), who enclosed vegetation in chambers ventilated with filtered or unfiltered air at the side of major roads in Switzerland, and found greatly increased aphid performance in unfiltered air, provides confirmation that these types of response can be caused by the direct effects of roadside air pollutants, most probably nitrogen oxides.

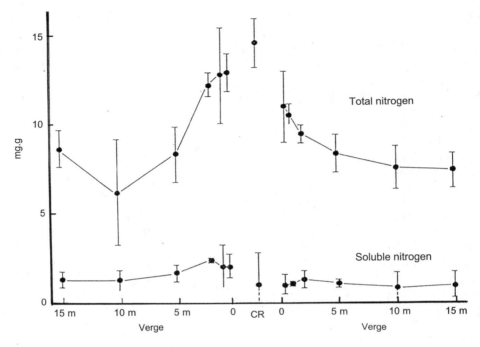

**Figure 4.** Effect of distance from the M1 on plant total and soluble nitrogen concentrations. From Port & Thompson (1980).

However, it should be noted that Braun & Flückiger (1984), in a parallel study, were able to show that spraying with salt solution at levels used along Swiss motorways also stimulated aphid population development and increased the host food quality. Hence, these responses of insect herbivores, which have been reported close to roads may have complex, multi-factor causes. It is also unclear from the literature to what extent altered plant-herbivore relationships might be found away from the immediate roadside, and what the longer-term ecological consequences of these altered plant-herbivore relationships might be.

## Tree growth and physiology

Studies of tree growth close to major roads offer the potential for examining the historical impact of motor traffic through the use of dendrochronological techniques. However, there have been few studies, which have used this approach. One preliminary report by Joos (1989) in Switzerland compared the historical growth patterns of spruce and beech trees in ecologically comparable sites within 100m, 1km and 2km from two roads with traffic densities of 35,000 and 45,000 vehicles per day, which were opened in the mid-1960s. In the case of beech, there was no evidence of an effect of proximity to the road. However, in spruce, the growth increments of trees closest to the roads appeared to have decreased since they were opened, in particular since the mid-1970s, when the trees close to the roads appeared less able to recover from the effects of drought years.

This is, however, only one limited piece of research and more studies of this nature would be very valuable in assessing the long-term effects of roads on adjacent woodland. It is also impossible to attribute the effects to specific factors associated with the road. While direct effects of air pollution may be involved, effects mediated through pollutants accumulated in the soils may be significant. For example, Majdi & Persson (1989) found that the fine root biomass and length of spruce trees was lower 20m from a major road than it was 200m away, an effect they attributed to lead concentrations being higher close to the road. Woodland nutrient cycling may also be influenced by elevated soil metal concentrations, through effects on microbial activity and rates of decomposition (Baath, 1989). For example, Cotrufo et al. (1995) found lower rates of decomposition of oak litter at an urban roadside site than at a rural control site.

## Effects on heathland species composition

The evidence outlined earlier that nitrogen deposition may be responsible for significant regional effects on vegetation across the UK suggests that the same type of effect might also be found on a more local scale close to major roads. However, with one recent exception, there have not been adequate empirical field studies in the UK to determine if this is the case. This exception concerns the changes from *Calluna*-dominated heathland to acid grassland, dominated by species such as *Molinia* and *Deschampsia*, which are a widespread phenomenon in the Netherlands; although nitrogen deposition may not be the only factor involved in these changes in vegetation, there is little doubt that it plays a significant role (Bobbink & Heil, 1993).

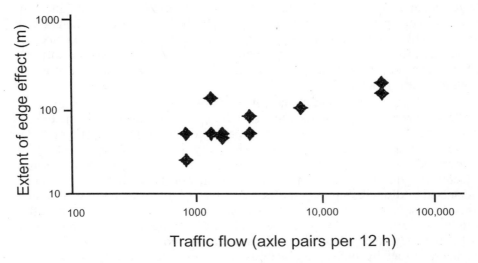

**Figure 5.** Relationship between traffic volume and extent of the edge effect, in terms of changes in heathland composition, in the New Forest. From Angold (1997).

Angold (1997) studied the effect of proximity to the A31 dual carriageway (12h traffic flow: 35,000 vehicles), and smaller roads with different vehicle densities, on the species composition of dry heathlands in the New Forest. Increased shoot growth and shoot nitrogen content of *Calluna* was found closer to roads, but cover of this species was somewhat lower nearer to roads. Increased leaf growth and cover of *Molinia* was also found close to roads and there was some evidence of increased invertebrate attack on this species close to roads. Multivariate analysis of species composition data showed significant correlations between the main axis of variation and distance from the road in almost all cases, with the axis being associated with increased performance of *Molinia* and decreased performance of *Cladonia* lichens.

Another observation of significance was the decreased cover and vigour of *Cladonia portentosa* lichens near to roads. Both the size of size of the lichen clumps, and the height of the stems, was decreased by about 40% at 25m from the A31 compared with 150m, although only the effect on stem height was statistically significant. Angold (1997) points out that there are a number of anecdotal reports that lichens may be eliminated along roadsides, possible due to vehicle exhaust pollution, but there is very little concrete field evidence to support these reports.

As with all such field studies, it is impossible to attribute the observed effects of proximity to the road to a specific factor. However, since the effects observed are very consistent with those observed in experimental studies of the impacts of enhanced nitrogen deposition on dry heathlands, it is probable that increased atmospheric concentrations of nitrogen oxides was the key causal factor (Angold, 1997). It is important to note that the botanical effects could be detected at distances up to 200m from the busiest road studied (the A31 dual carriageway). Angold was able to demonstrate a relationship between traffic flow and the extent of the 'edge effect', in terms of the distance from the road at which significant botanical effects could be observed, using subjective criteria verified by multivariate analysis of species cover data (Figure 5).

This empirical approach could in theory be used in assessing the extent of the possible impacts of new road schemes, although it would need more extensive validation, and would be different for different types of community. If this type of empirical relationship, based on careful and methodological field observation, could be linked to modelling of the concentrations of specific pollutants over the distances at which field effects were detected, a powerful predictive tool could potentially be developed.

## CONCLUSIONS

The key conclusions of this paper can be summarised as follows:-

- There is good empirical evidence that air pollutants, along with other factors linked to roads and traffic, significantly influence roadside ecology, although more detailed studies to quantify these impacts and relate them to specific pollutant combinations are needed.

- Air pollutants from road traffic make a significant contribution to regional acidification, eutrophication and photochemical ozone production, all of which are known to have significant ecological impacts in the UK. National policy has recognised these problems, and policies have been developed to reduce national emissions of $SO_2$, $NO_2$ and VOCs.

- However, the effects of air pollutants on the ecology of areas at distances of 10m to 500m from a major road are uncertain; given that the greatest concerns are likely to relate to impacts on sites of high nature conservation which are close to major roads, rather than the unique and artificial roadside habitat, more studies are urgently needed to define the scale and significance of these effects.

- There is evidence of increased pollutant emissions from road traffic associated with visitor access to sensitive areas, such as national parks. The impacts of the increased local pollutant concentrations arising from these emissions are uncertain and need further evaluation to inform the planning process in these areas.

- Critical levels and loads provide a useful tool for assessing the potential impacts of road developments. However, they relate to individual pollutants, and their application to roadside situations is limited by the fact that the impacts of the specific pollutant mix found in such locations are poorly defined. More empirical approaches based on field studies, using distance from road and vehicle numbers as predictive parameters, could usefully be developed as a planning guide to complement the use of dispersion modelling linked to air quality criteria.

- Almost no studies to date have considered the effect of time, as well as proximity to a road, in assessing ecological impacts. Since cumulative effects of deposition of pollutants such as metals and nitrogen oxides are to be expected, but have not been studied, and since emissions from any new road scheme are likely to continue for decades, it is impossible from current evidence to evaluate the long-term ecological impacts. This is an urgent research priority for the future.

- Although there is limited evidence from field studies of local impacts of roads on sensitive plant and animal communities, the potential for cumulative effects over time mean that a precautionary approach is essential in planning new road developments.

## REFERENCES

**Angold PG. 1997.** The impact of a road upon adjacent heathland vegetation: effects on plant species composition. *J appl. Ecol.* **34:** 409-417.

**Ash CPJ, Lee DL. 1980.** Lead, cadmium, copper and iron in earthworms from roadside sites. *Environ. Pollut. Series A.* **22:** 59-67.

**Ashmore MR. 1997.** Pollution. In: Crawley MJ. ed. *Plant Ecology.* Oxford.: Blackwell Scientific. 568-581.

**Ashmore MR. Ainsworth N. 1995.** Effects of ozone and cutting on the species composition of artificial grasslands. *Funct. Ecol.* **9:** 708-712.

**Ashmore MR. Davison AW. 1996.** Towards a critical level of ozone for natural vegetation. In: Karenlampi L, Skarby L. eds. *Critical Levels for O₃ in Europe.* Kuopio: University of Kuopio. 58-71.

**Atkins DP, Trueman IC, Clarke CB. 1982.** The evolution of lead tolerance by *Festuca rubra* on a motorway verge. *Environ. Pollut. (Series A).* **27:** 233-241.

**Baath E. 1989.** Effects of heavy metals in soil on microbial processes and populations. *Water Air Soil Pollut.* **47:** 335-379.

**Bell JNB ed. 1994.** The ecological effects of increased aerial deposition of nitrogen. British Ecological Society, *Ecological Issues no 5.* Preston Mountford:  Field Studies Council.

**Bell JNB, Ashmore MR, Wilson GB. 1991.** Ecological genetics and chemical modifications of the environment. In: Taylor Jr. GE, Pitelka GE, Clegg, MT. eds. *Ecological Genetics and Air Pollution.* New York: Springer-Verlag. 33-59.

**Bell S, Ashenden TW. 1997.** Spatial and temporal variation in nitrogen dioxide pollution adjacent to rural roads. *Water Air Soil Pollut.* **95:** 87-98.

**Bobbink R, Heil GW. 1993.** Atmospheric deposition of sulphur and N in a heathland ecosystem. In: Aerts R, Heil GW. eds. *Heathlands: Patterns and Processes in a Changing Environment.* Dordrecht: Kluwer Academic. 25-50.

**Braun S, Flückiger W. 1984.** Increased population of the aphid *Aphis pomi* at a motorway: Part 2 - The effect of drought and deicing salt. *Environ. Pollut. (Series A).* **36:** 261-270.

**Braun S, Flückiger W. 1985.** Increased population of the aphid *Aphis pomi* at a motorway: Part 3 - The effect of exhaust gases. *Environ. Pollut. (Series A).* **39:** 183-192.

**Clark AI, McIntyre AE, Reynolds GL, Kirk PW, Lester JN, Perry R. 1988.** Statistical analysis of gaseous air pollutant concentrations at urban, rural and motorway locations. *Environ. Technol. Letters.* **9:** 1303-1312.

**Cotrufo MF, de Santo AV, Alfani A, Bartoli G, de Cristofaro A. 1995.** Effects of urban heavy metal pollution on organic matter decomposition in *Quercus ilex* L woods. *Environ. Pollut.* **89:** 81-87.

**Farmer AM. 1993.** The effects of dust on vegetation - a review. *Environ. Pollut.* **79:** 63-75.

**Flückiger W, Oeetli JJ, Flückiger-Keller H, Braun S. 1970.** Premature senescence in plants along a motorway. *Environ. Pollut.* **18:** 171-176.

**Freer-Smith PH, Holloway S, Goodman A. 1997.** The uptake of particulates by an urban woodland: site description and particulate composition. *Environ. Pollut.* **95:** 27-35.

**Grue CE, Hoffman DJ, Bayer WN, Franson LP. 1986.** Lead consentrations and reproductive success in European starlings *Sturnus vulgaris* nesting within highway roadside verges. *Environmental Pollution. (Series A).* **42:** 157-182.

**Hill S. 1990.** Some indirect ecological effects of road schemes. [MSc Thesis] London: Imperial College Centre for Environmental Technology.

**Ieradi LA, Cristaldi M, Mascanzoni D, Cardarelli E, Grossi R, Campanella L. 1996.** Genetic damage in urban mice exposed to traffic pollution. *Environ. Pollut.* **92:** 323-328.

**INDITE. 1994.** *Impacts of Nitrogen Deposition in Terrestrial Ecosystems.* London: Department of the Environment.

**Jefferies DJ, French MC. 1972.** Lead concentrations in small mammals trapped on roadside verges and field sites. *Environ. Pollut.* **3:** 147-156.

**Joos KA. 1989.** Investigation of a possible direct influence of highway traffic on nearby woods. In: Bucher JB, Bucher-Wallin I. eds. *Air pollution and Forest Decline.* 436-438.

**Kingston L, Leharne S, McPhee E. 1988.** A survey of vehicular lead deposition in a woodland ecosystem. *Water Air Soil Pollution.* **38:** 239-250.

**Majdi H, Persson H. 1989.** Effects of road-traffic pollutants (lead and cadmium) on tree fine-roots along a motor road. *Plant & Soil.* **119:** 1-5.

**Port GR & Thompson JR. 1980.** Outbreaks of insect herbivores on plants along motorways in the United Kingdom. *Journal of Applied Ecology.* **17:** 649-656.

**Sawada S, Hazama Y, Hayakama T, Totsuka T. 1989.** Simulation of effects of atmospheric ethylene on tree leaf shedding. *Environ. Pollut.* **61:** 173-185.

**Sjodin A, Ahlfors A, Starby L, Grennfelt P. 1988.** Effects of vehicle exhausts near roads. *Report 3401, Swedish Natural Environmental Protection Board.*

**Spencer HJ, Scott NE, Port GR, Davison AW. 1988.** Effects of roadside conditions on plants and insects. I. Atmospheric conditions. *Journal of Applied Ecology.* **25:** 699-707.

**Spencer HJ, Port GR. 1988.** Effects of roadside conditions on plants and insects. II Soil conditions. *J appl. Ecol.* **25:** 709-715.

**Taylor HJ, Ashmore MR, Bell, JNB. 1986.** *Air Pollution Injury to Vegetation.* London: Intitute of Environmental Health Officers.

**UK CLAG. 1997.** *Critical Levels of Air Pollutants for the United Kingdom.* Sub-group Report on Critical Levels of Critical Loads Advisory Group. London: Department of the Environment.

**UK PORG. 1997.** *Ozone in the United Kingdom 1995.* 4th Report of the Photooxidant Review Group. London: Department of the Environment.

**Wade KJ, Flanagan JT, Currie A, Curtis DJ. 1980.** Roadside gradients of lead and zinc concentrations in surface-dwelling invertebrates. *Environ. Pollut. (Series B).* **1:** 87-93.

**Williamson P, Evans PR. 1972.** Lead levels in roadside invertebrates and small mammals. *Bull. Environ. Contam. Toxicol.* **8:** 280-288.

**Woolgrove CE, Woodin SJ. 1996.** Current and historical relationships between tissue nitrogen content of a snowbed bryophyte and nitrogenous pollution. *Environ. Pollut.* **91:** 283-288.

**Wu L, Antonovics J. 1976.** Experimental ecological genetics in *Plantago.* II. Lead tolerance in *Plantago lanceolata* and *Cynodon dactylon* from a roadside. *Ecology.* **57:** 205-208.

# 13

# The ecological effects of road lighting

A R OUTEN

Life on earth has evolved in the company of alternating daylight and darkness. Virtually every group of living organisms makes use of this daily cycle and the seasonal changes in day length as sources of environmental information. Diurnal rhythms are extremely important components of animal behaviour and day-length also influences circannual activities of plants and animals. Geographical distribution, seasonal biology, growth, metabolism and behaviour of living things are profoundly influenced by the length of the photoperiod.

This is understandable, as light is normally the most predictable factor of the physical environment. In contrast to temperature 'natural' day length is always the same each year for a given season and latitude. As has been known since the 1940s, it is in fact more strictly 'the period of uninterrupted darkness' that governs much of the daily and annual behaviour of organisms and it is a well-established scientific principle that artificial disruption of this can induce unnatural activity in plants and animals.

The former predictability of light so fundamental to life is no longer reliable. Outdoor lighting has transformed the nocturnal face of the earth. Individual countries and large towns can be identified from space by their nocturnal illumination. In the USA the distribution of outdoor lighting can be seen to coincide with that of the country's population with illumination clustered around all large metropolitan areas, and the greatest concentration in the northeast corridor (Croft, 1978).

Recently astronomers have begun to express their concern at difficulties in carrying out their observations due to the increasing level of background illumination of the night sky. Electric lighting in some areas has increased nocturnal sky brightness as much as 20-fold (Hendry, 1984). Modifications of the natural environment on such a dramatic scale are unlikely to be without significance to flora and fauna and there is now increasing evidence that outdoor lighting is having a wide diversity of effects on living things. In a small, densely populated and heavily urbanised country such as Britain, this potential problem is particularly acute.

Literature on the ecological effects of artificial light in natural environments is sparse. Much of the research that has been done is theoretical or has been directed towards food production or pest control. The following account aims to provide a synthesis of some of the work demonstrating the impact of artificial light.

## PLANTS

In plants the breaking of bud dormancy, flowering, leaf fall and in some cases seed germination are controlled by day-length. 'Short day plants' (e.g. common toadflax *Linaria vulgaris*, Chiltern gentian *Gentianella germanica*) flower only as the period of uninterrupted darkness increases. 'Long day plants' (e.g. wild daffodil *Narcissus pseudonarcissus*, wood anemone *Anemone nemorosa*.) flower when the duration of the dark period decreases below a critical value. Day neutral plants (including many common weed species e.g. dandelion *Taraxacum* Sect. *vulgaria*) are unaffected by day-length.

Under controlled conditions a short day plant given a brief flash of light during the night does not flower. Conversely breaking up the long nights with brief flashes of light induces long day plants to flower (Reid, 1969), a principle utilised by commercial growers to produce flowering chrysanthemums (*Chrysanthemum* spp.), and other plants out of season.

As in most ecological aspects of this subject the situation is complex but it has been suggested that night length is measured so accurately by plants, that some short-day plants will not flower if the night is even one minute shorter than the critical length (Campbell, 1990). Differences in natural daylength between the extreme south and extreme north of Britain are sufficient to affect flowering of our native plants (Turrill, 1948).

Under laboratory conditions harmful effects of sodium vapour lighting on plants, through disruption of photoperiodic regulation of growth and development, have been shown by several workers (e.g. Sinnadurai, 1981; Cathey & Campbell, 1975; Shropshire, 1977). However Andresen (1978), who poses the intriguing question *"Do street lights turn city trees into late-growing insomniacs?"* suggests the effects are apparently greater under greenhouse conditions than outdoors. So far as I have been able to determine no studies have been carried out on possible changes in natural vegetation structure and composition as a consequence of the effects of artificial lighting, though these should certainly be expected.

## MAMMALS

The cover of darkness is important to many mammal species in the avoidance of predation though for some species light is important for foraging activity. In North America it has been shown that Oldfield mouse *Peromiscus polionotus*, show 70% activity at full moon but only 23% activity at quarter moon. (Wolfe & Summerlin, 1989). It has been suggested, that hedgehogs (Erinaceidae) and shrews (Soricidae) may be attracted to lit areas of roads in their foraging for invertebrates and hence so often fall victim to traffic (Molenaar *et al.,* 1997).

If caught in headlamps species such as hedgehogs *Erinaceus europaeus*, hares *Lepus capensis* and rabbits *Oryctolagus cuniculus* tend to either freeze, crouch low to the ground or run within the lit area (especially hares) such that they run a particularly high risk of being hit. Yellow headlights (which are more effective through fog than white light) have been shown to be much less hazardous in this regard than white ones (Schachinger, 1962).

Some nocturnal mammals are likely to be disturbed by the presence of bright illumination and could be deterred from using established foraging areas etc. This is a further pressure on species such as badger *Meles meles* and otter *Lutra lutra* for which there is existing statutory protection though it is unlikely existing legislation would be applicable in this situation. Lighting of motorways might have a positive safety benefit in deterring deer, badgers etc., (Gallagher *et al.* in Reed, 1981) however where preference crossing points are subsequently illuminated they continue to be used (Rumar, in Reed, 1981) Lighting alongside river corridors, or adjacent to foraging areas or open countryside could also be detrimental.

Much has been made of the benefit to bats of the feeding opportunities provided by insects attracted to street lighting. However, recent work by Rydell & Racey (1993) has shown that although species such as *Nyctalus, Vespertillio, Eptesicus, Pipistrellus* do take advantage of this, such locations are not exploited by the slower flying species (e.g. *Myotis, Plecotus* and *Rhinolophus*) which shun bright lights, include most of those particularly vulnerable in Europe. High densities of bats were only detected near white (mercury) street lamps. Fenton & Rautenbach (1986) consider that horseshoe bats *Rhinolophus* sp., and long-eared bats *Plecotus* sp., shun bright lights as a predator avoidance strategy. Hence the effect of lighting on bat populations might be beneficial to some species but detrimental to others and the net result may well be an altering of the balance between species.

The species, which feed at road lights, are however the most vulnerable as potential traffic victims. Low flying species are particularly susceptible.

Pregnant females and juveniles form a significant proportion of such casualties (Rydell, 1991).

In most bat species there is an evening period of activity lasting up to about two hours followed by another before dawn. These two flights correlate with the peak flight times of nocturnal insect prey. Studies suggest that pipistrelle bats *Pipistrellus pipistrellus* leave their roosts when the light intensity reaches a critical level after sunset and that time of emergence is unaffected by vagaries of the weather (Swift, 1980).

It has been suggested that the nocturnal restriction of bat activity is a predator avoidance strategy and that larger bats are most likely to avoid light because of their lower manoeuvrability (Speakman, 1991) whilst 'lunar phobia' has been demonstrated in one neotropical fruit bat species (Morrison, 1978). It has been shown that bats which feed over water e.g. Daubenton's bat *Myotis daubentonii*, and serotine *Eptesicus serotinus*, are deterred by illumination (Reinhold, 1993).

Limpens (*pers. comm.*) has suggested that continuous lighting along roads creates barriers which bats will not cross. He advises leaving stretches of unlit road to avoid isolation of bats and this has already been implemented on some motorways in Holland and Sweden. A similar situation has also been noted by Rieger in Switzerland. Dr. Yalden of University of Manchester has suggested (*pers. comm.*) that bat colonies would probably desert roost sites that became illuminated and states that lights in a roof space have been suggested as a humane deterrent. English Nature has also apparently given this advice (Jones *pers. comm.*).

It has been demonstrated that in at least one species of bat, daylength influences the male reproductive cycle (Beasley & Zucker, 1984) and the timing and level of peak body weight through autumn fat deposition (Beasley, Peltz & Zucker, 1984). Molenaar *et al.* (1997) suggest that mammal species nesting above ground such as harvest mouse *Micromys minutus*, dormouse *Muscardinus avellanarius*, and squirrels *Sciurus carolinensis* are most susceptible to the influence of illumination on their reproduction. Physiological and hormonal changes initiating hibernation in creatures such as dormice and bats have also been linked to day-length.

## BIRDS

For birds in the temperate region night-length initiates such behaviour as courtship and mating, reproductive cycles, migration, moulting and seasonal changes in plumage.

As long ago as 1923 it was demonstrated that dawn singing of thrushes began within a minute of a critical light level being reached (Rawson, 1923).

This again indicates the degree of precision involved in the control of behaviour by light. The Royal Society for Protection of Birds report receiving many calls of robins *Erithacus rubecula* singing as early as January and in the middle of the night (*pers. comm.*). In a recent song by Smith & Sergeant (in the mould of Flanders & Swann) they give an account, to the familiar tune, of how *"Twas a Robin that Sang in Berkeley Square"*. Other birds which have been reported in literature as singing at night under the influence of artificial lighting include bluethroat *Luscinia svecica*, reed bunting *Emberiza schoeniclus*, chiffchaff *Phylloscopus collybita*, dunnock *Prunella modularis*, blackbird *Turdus merula* and nightingale *Luscinia megarhynchos*.

Farner (1964) has demonstrated the photoperiodic control of reproduction in birds and that artificially increasing day length will induce hormonal, physiological and behavioural changes initiating breeding. Researchers have found that about 60 wild bird species can be brought into premature breeding condition by experimental exposure to artificially short nights in winter. None failed to respond though effects are more pronounced in males (Lofts & Merton, 1968).

Lack (1965) noted that robins and blackbirds began laying one or two weeks earlier in gardens than in woods and also stopped breeding later. He was unable to suggest an explanation for this but it could perhaps be partly due to the effects of artificial lighting influencing the birds in the urban environment. Premature breeding in the wild could result in brood loss due to cold weather or non-availability of food insects and consequently wasted energy.

The attraction of birds to lights has been known for a long time particularly to lighthouses (Clarke, 1912; Axell, 1964; Elkins, 1983). Ceilometer lights at airports caused heavy mortality until they were modified by shifting their spectra into the ultraviolet and turning them on only briefly (Howell *et al.,* 1954; Laskey, 1956; Terres, 1956).

It has also been estimated that as many as one million migrants die at illuminated television towers annually (Aldrich *et al.,* 1966). In one incident lights at a fishing vessel attracted 1.5 metric tons of crested auklets *Aethia cristatella,* severely endangering the stability of the vessel (Dick & Donaldson, 1978).

Studies of bird kills at lighthouses and other tall structures in United States and Holland demonstrated that the greatest number of mortalities occurred around new moon nights Verheijen (1980, 1981). Artificial light in moonless conditions impairs the orientational ability of birds. In the presence of moonlight, whether or not weakened and scattered by clouds, the birds are not disorientated by artificial light sources.

Mead (1983) suggests that snipe *Gallinago gallinago*, water rail *Rallus aquaticus*, sedge warblers *Acrocephalus schoenobaenus* and reed warblers *A. scirpaceus* are particularly notorious victims of attraction to tall lighted structures. In general passerines which fly at lower altitudes are more susceptible than the stronger flying waders, ducks and geese which fly at higher altitudes.

Reed *et al.* (1985) studied the attraction of artificial lighting to procellariform birds (shearwaters and petrels) in Hawaii during Autumn. Street Lights from the largest coastal resorts were shielded on alternate nights to prevent upward illumination during two fledging seasons. Most attraction was shown to occur 1-4 hours after sunset and a full moon dramatically decreased the attraction of artificial lights. Attraction was decreased by up to 40% by shielding.

Telfer *et al.* (1987) again noted a significant reduction in the attraction of Hawaiian seabirds to artificial lighting at full moon and found fledglings particularly susceptible after first leaving the nest. Attraction and disorientation was most pronounced in urban coastal areas particularly river mouths. It is important to appreciate however that light attraction is a general problem in birds, not restricted to petrels or even seabirds (Reed *et al.*, 1985). Hill (1992) suggests shielding of streetlights should be incorporated into construction of new roads which are close to bird breeding areas.

It has been found that some birds can even be temporarily blinded by road lighting (Pettingill, 1970). In Texas hundreds of dead songbirds have been found beneath lamp posts every spring (James, 1956).

Nocturnal birds are likely to be disturbed by the presence of bright illumination. As many of these species are already under threat this is a further pressure on remaining populations. Barn owls *Tyto alba*, and nightjars *Caprimulgus europaeus* are both Red Data Book species known to be generally susceptible to human disturbance. British populations of barn owls and long-eared owls *Asio otus* (our most nocturnal species) have declined much more dramatically than tawny owls *Strix aluco* and little owls *Athene noctua* (the least nocturnal and which until comparatively recently were considered to be increasing). No studies have been done to determine whether there is any correlation between these changes and increased nocturnal illumination.

Artificial lighting has the potential to provide more feeding time for birds by enabling nocturnal feeding. However increasing feeding time for birds in this way could have a detrimental effect on a prey population leading ultimately to food shortage for the birds and also creates the potential for an unnatural competition in niche (e.g. diurnal raptors feeding at night rather than those which specifically evolved for this role).

The extended feeding time for Bewick's swans *Cygnus columbianus*, feeding under illumination at Slimbridge, Gloucs. means that fat is laid down more quickly and they can hence reach migration condition much quicker. Early departure, exacerbated by the trigger of artificially extended daylength, could result in birds arriving too early in their breeding habitats with damaging consequences (Rees, 1982).

Hill (1992) reports incidents of waders feeding at night under the artificial illumination provided by street lighting. Amies (1990) suggests Herons were utilising motorway lighting to facilitate night feeding. Other species which have been reported using artificial light for nocturnal feeding include robins (Green, 1978; England, 1978), blackbirds (England, 1978), swallows *Hirundo rustica* (Knox, 1990); alpine swifts *Apus melba* in Switzerland (Freeman, 1981) etc. Kestrels *Falco tinnunculus* can be observed hunting at night under artificial light on motorway verges.

## REPTILES AND AMPHIBIANS

Although day length is undoubtedly important in annual activities including reproduction, moulting and hibernation there is no evidence of direct attraction or repulsion in most reptiles.

Outside Britain geckoes will feed on insects attracted to artificial light. Newly hatched marine turtles are disorientated by artificial lighting from roads where beaches receive illumination (e.g. McFarlane, 1963). Adult green turtles *Chelonia mydas,* avoid laying on beaches which are illuminated even when the lights are not aimed directly at the beach (Carr *et al.,* in Verheijen 1985). Amphibians are far more susceptible to light attraction than reptiles. In one study covering 121 species of frogs and toads, 87 were found to exhibit a positive light reaction. Toads are especially sensitive while some amphibians avoid light (Jaeger & Hailman, 1973). Aggregations of common toad *Bufo bufo*, have been reported under street lamps increasing their vulnerability as traffic victims (Baker, 1990). The non-British green toad *Bufo viridis*, is reported to show a preference for feeding under artificial light (Balassina, 1984).

## FISH

As in many cases little work has been done, yet as long ago as 1950 it was demonstrated that artificially increasing day length would stimulate brook trout *Salvelinus fontinalis* into breeding up to four months early (Hazard

& Eddy, 1950). As with any species the consequence of breeding out of season can result in high mortality of progeny and of weakened adult individuals. The significance of daylength in the feeding and other activity of a number of species has been demonstrated (Cui *et al.,* 1991; Glass, Wardle & Mojsiewicz, 1986; Glass & Wardle, 1989) and in 1996 an Aberdeen angling club argued successfully in Court that a tennis club which had introduced floodlighting to their courts had had a detrimental effect on fish catches.

## INSECTS

Moths and other night-flying insects are attracted to lights and concern has been expressed that street lighting  may have affected their populations in some way. However despite universal awareness that electric light disturbs behaviour of nocturnal insects, the ecological impact of outdoor lighting has never been comprehensively assessed or studied, a conspicuous dilemma applicable to all groups of organisms. A review of the potential impacts of lighting on moths from available literature was undertaken by Frank (1988). (Table 1.)

Although nocturnal Lepidoptera (Moths) are the subject of most attention in their response to light, Insects from a wide range of orders are known to be attracted to light sources including Neuroptera (lacewings), Coleoptera (beetles), Hemiptera (bugs); Trichoptera (caddis flies); Diptera (two-winged flies); Hymenoptera (principally wasps and some ichneumons) and Orthoptera (especially bush crickets).

In fact the generalised statement that moths are attracted to light is to some extent misleading. As the Robinson brothers demonstrated in 1950 a high general level of illumination may cause night-flying insects to cease flying and settle, just as they do at daybreak. They noted that a single light source isolated in a dark area was the most effective in attracting insects. That is why insects fly towards a lamp at night but when they arrive in the lit area they settle, preferably into a nearby shadow. A light seen at a great distance (e.g. the sun or moon) enables an insect to fly in a straight line and thus maintain a constant angle with the direction of the light. Artificial lights mislead an insect in its attempts to navigate in this way. As might be expected artificial light attraction varies with the lunar cycle and is lowest at full moon.

Robinson (1952) distinguished between the 'power' of the lamp (measured in lumens) and the 'surface brilliance' (measured in lumens per sq.ft. of the lamp surface). He states that as the power was increased more individual insects were caught in a light trap whereas an increase in surface brilliance (e.g. a smaller lamp giving out the same total light) brought in an increased number of species.

**Table 1.** Types of street lighting and their effects on insects. Adapted from Frank (1988).

| Low pressure sodium lamps | Yellow lamps of most roadsides. | Emit nearly all light at one wavelength to which we are particularly sensitive so appear bright to us but with a low overall level of brightness | Low attraction to insects and do not disturb circadian rhythms. |
|---|---|---|---|
| High pressure sodium lamps | Appear pinkish- yellow to us. | Typically 2.5 times as bright as low pressure sodium lamps and emit most light over a broad band of long wavelengths,. | Attract most insects |
| Mercury Lamps | Appear bluish-white to us. | Emit light more or less evenly over a very broad spectrum. About 1.5 times as bright as low pressure sodium lamps | Attract most insects in large numbers (due to presence of Ultra-violet light to which insect eyes are particularly sensitive). |

Although accounts are contradictory in assessing the distance from which moths are attracted to light the experimental methods of some of these estimates have been criticised. Stewart *et al.* (1969) reports positive response to a 15W actinic tube at between 60 and 200 metres depending on the moth species though others have questioned this (e.g. Baker & Sadovy, 1978; McGeachie, 1988).

Flight Suppression in moth and other insects by a high general level of illumination is also not straightforward. Whereas some may remain quiescent for only a moment others may remain so for the entire night (Blest, 1963; Graham *et al.*, 1964). In other cases quiescent moths may be excited into flight (Collins, 1934; Hsiao, 1972). It is widely known that day flying moths are also sometimes attracted to light at night, though it is unclear whether the flight is initiated by artificial lighting.

Electric lights may also actually divert moths away from them (Robinson & Robinson, 1950; Nomura, 1969; Hsiao, 1972) an effect that may depend in part on the spectral composition of the light (Nomura, 1969). Evidence that moths avoid large illuminated areas is however inconclusive though this behaviour is more difficult to demonstrate than light attraction (Nomura 1969).

The same urban changes that increase outdoor lighting also lead to a fragmentation of habitats and the result is the creation of small isolated relict insect colonies exposed to illumination. Urbanisation increases both vulnerability and exposure of such insect populations to artificial lighting. Frank (1988) suggests that as habitats become isolated by urban sprawl, moths may have to traverse dozens of kilometres of densely illuminated territory to arrive at breeding grounds. In addition he quotes some evidence that lighting along roads following topographic features such as valleys, river corridors and coastlines might selectively interfere with moth migrations at least in USA.

Over three hundred and fifty species of larger moths (about a third of the British species) have been collected at a single light trap in England (Williams, 1939) and comparable findings are also reported (Robinson & Robinson, 1950; Bretherton, 1954; Langmaid, 1959; Hosney, 1959; Waring, 1994). Thousands of moths have been known to fly to a single lamp in one evening (Robinson & Robinson, 1950; Nau, et al,. 1987; Webb & Malt, 1977; Feltwell, 1996) and huge aggregations have been reported around urban light sources (Howe, 1959). However certain species of nocturnal moths are rarely attracted to lights (Bretherton, 1954; Taylor & Carter, 1961; Skinner, 1984). In some species males are more frequently attracted than females (Skinner, 1984).

Physiological, behavioural and environmental factors may all be involved in determining which species of moths fly to light and when (Geier, 1960). In addition artificial light attraction varies with the lunar cycle and is lowest at full moon (Williams et al., 1956; etc.). Outdoor lighting may also act selectively on particular individuals within a population, perhaps selecting against those most strongly attracted to light for example. Frank (1988) suggests that the fact that some species of moth are not attracted to light sources might indicate that behavioural modification of the flight to light response has already occurred.

Illumination by these sources can cause congregation of light attracted insects in the vicinity. Here they could be an easy prey for bats, or the following morning to birds. In locations where Rothamsted mercury vapour moth traps are used nightly it is well known that birds (particularly blackbirds, robins and tits) will establish a routine of coming to feed in the morning on moths settling nearby (e.g. Waring, 1994).

Different species of moths fly at different times of the year and in some cases the flight period for the species may be only two or three weeks. Different species also fly at particular times during the night e.g. The garden tiger moth *Arctia caja,* rarely flies before midnight after which conditions are naturally darkest (Waring 1994).

Continuous usage of such bright light sources as floodlighting and security lighting could pose even greater potential problems for such species than normal street lighting in that the floodlights are generally of greater intensity and security lights are left on for extended periods, whereas in many areas street lighting is often timed to go off between 12pm. and 2am. Frank (1988) presents evidence that artificial light might result in a shift in flight period. Advancement or delay of flight times could disturb natural species segregation mechanisms.

In North America Lepidopterists have blamed light pollution for the decline in moth populations (Worth & Muller, 1979; Pyle *et al.,* 1981; etc.). There is a perception among some Entomologists that urban locations in Britain support a far lower diversity of moth species than 30 - 40 years ago even where there has been little change in the composition of their vegetation and hence availability of larval food plants. Woiwood & Riley (1992) report a notable decrease in the numbers of many species and an overall fall in moth populations of over 60% over the last fifty years. Apparent declines in moth populations and species diversity are discernible even where vegetation does not seem to have significantly altered. Although there are other possible causes it may be that increased lighting has been a factor in this apparent deterioration in the moth fauna.

Frank (1988) cites evidence of decrease in moths at urban lamps but considers many factors may contribute to this including: decline in moth populations, dilution of moths among the multitude of urban light sources and suppression of the flight to light response as a result of overall increase in background illumination. It is also conceivable that species composition might be affected in densely illuminated urban areas by the lighting selectively favouring day-flying species or those which are not attracted to light or those that do not fly at all. Species with flightless females include mottled umber *Erannis defoliaria*, early moth *Theria primaria*, vapourer *Orgyria antiqua* etc.

Other sources suggest no downward trends in moth populations are apparent (despite wide fluctuations from year to year) and that moth populations in areas undergoing urban changes can substantially recover despite electric lighting (Taylor *et al.,* 1978). Some American attempts to exploit light traps to control pest species have failed (Cantelo, 1974; Hienton, 1974).

However it is generally the case with common species of any group of organism that they are more tolerant and resilient of such pressures in their environment  Insect pests and perennial garden weeds can be notoriously difficult to eliminate for precisely this reason whereas more desirable species, more specialised in their requirements, are readily exterminated.

Other causes have been suggested for declines and extinction of individual species particularly agricultural changes, loss of habitat and changing climate (Bretherton, 1951; Ford, 1972; Heath, 1974). However studies on *Hydraecia petasites* (The butterbur moth) a scarce species in Finland suggest that continuous light trapping could destroy the population. This species does occur in Britain though it is infrequent. As the authors point out this species is only mildly attracted to light so the effect might conceivably be more severe for other moths (Vaisanen & Hublin, 1983).

Bright light can lower sensitivity of moth eyes 1000 fold (Bernhard & Ottoson, 1960; Eguchi & Horikushi, 1984 etc. as quoted by Frank, 1988). If the moth remains at the lamp and then flies away, full visual sensitivity may not return for 30 min. or longer (Bernhard & Ottoson, 1960; Agee, 1972). A moth flying away from a lamp into relative darkness on a cloudy moonless night may in effect be functionally blind until enough time has elapsed for it to become fully dark-adapted.

Oviposition (egg laying) by moths can also be disturbed by lighting. Although males usually outnumber females at light traps, the vast majority of female moths collected at light are gravid. 'Flight to light' can shift oviposition to sites located near the lamp and eggs may be deposited on lampposts, windows, buildings and other unsuitable sites in the vicinity of the light source. Egg densities may be several times higher on plants near lights leading to possible overcrowding and thus enhancing the risk of starvation, predation and disease. It is probable that the lighting diverts the ovipositing females rather than by stimulating oviposition. In *Ostrinia nubilalis* (The European cornborer) it has been found that in the laboratory nocturnal illumination suppresses egg laying but in cornfields the females will lay near lamps (Beaty *et al.,* 1951; Skopik & Takeda, 1980). Similar observations have been reported in other species whilst in further species outdoor lighting appears to decrease egg laying (Frank, 1988). Such anomalies demonstrate the difficulty in extrapolating data from individual species and in interpreting laboratory data in a natural context.

It has been suggested that lamps might disturb oviposition which is synchronised to lunar rhythms (Nemec, 1969, 1971). Frank (1988) considers that the extent of possible correlation between moth activity and lunar cycles, together with possible disturbance of this by artificial lighting, warrants investigation.

It has been shown that light may suppress mating in some species and that under laboratory conditions this can occur even at levels of light well below the general level of outdoor electric lighting. In one study male hawkmoths attracted to light traps baited with virgin females failed to seek out the females (Frank, 1988). Several entomologists have reported failure of hawkmoths to mate unless in darkness (*pers. comm.*). In one moth species electric light has also been shown to suppress release of sex pheromones in females and male response to these (Fatzinger, 1979). In another species a minimum of at least 7 hrs. relative darkness is necessary to induce mating (Lukefahr & Griffin, 1957). However studies on some other moth species suggest their mating is not inhibited by lighting (Frank, 1988 from lit.).

There have been no detailed investigations into the possible effects on timing of nocturnal behaviour in moths attracted to light. Frank (1988) questions whether the mating period of a male moth exposed to light would still coincide with a female not so exposed; or whether pheromone release in a light attracted female would still correlate with the male flight period. Shifts in mating times might cause sympatric, closely related species to attempt to mate. One reason why such matings would not normally occur is because they are temporally segregated by different mating periods (Tuttle, 1985).

Friedrich (1986) states that the fundamental factor required to induce continuous development in hawkmoths is change in photo-periodicity. Additionally in some cases fertile moths only result from larvae kept under short-day conditions.

Heinig (1982) has carried out intensive studies on hummingbird hawkmoths *Macroglossum stellatarum*, a migrant species to Britain. He concludes that in the Autumn the short day length (12 hours and less) stops the production of juvenile hormone and induces the adult to diapause with a delay in development of the gonads. (Diapause is a mechanism for dormancy during periods of unfavourable environmental conditions e.g. winter, through a state of suppressed development). In the spring, under the influence of increasing day-length, the hormone system increasingly secretes more juvenile hormone and the adults become sexually mature.

Most detailed studies on moth physiology have been on hawkmoths because of their large size but many other moth and butterfly families (e.g. Pieridae, tussock moths; Lymantridae, tiger moths and relatives: Arctiidae; and 'bombycids' including Drepanidae; Thyatiridae; Notodontidae; Thaumetopoeidae; Lymantridae and Nolidae) similarly respond to day length as the inducement to diapause and hence control of number of generations per year (Friedrich, 1986).

Within the Noctuidae (the largest family of moths in Britain) members of the subfamily Acronictinae (daggers and their relatives) show continuous development in artificially extended day length, but without a diapause (the species normally overwinter as a pupa) mating very often no longer takes place (Friedrich, 1986).

If as a consequence of artificial lighting moth larvae are not induced to produce overwintering pupae then the likely outcome for the resulting next generation is either that they will be killed by frost or indirectly as a result of non-availability of food plants following frosts. Even if such deaths do not result then there remains the possibility that the effects of artificial lighting could interfere with the reproductive hormones of the insects resulting in infertility. In some cases nocturnal illumination has been shown to interfere with feeding of adult moths (Herms, 1932; Brown. 1976) and this has been utilised in certain cases in effectively reducing crop damage by pest species (Nomura. 1969). In some other moth species it has been shown that illumination does not apparently interfere with feeding (Frank, 1988). In addition as many moths do not feed as adults (Skinner, 1984) this problem does not affect them.

Little work has been done on the impact of artificial light in the natural environment on other insect groups though it is suggested that outdoor lighting is potentially disruptive to the behaviour of a large mayfly living in the Thames (Bratton, 1990). Crowson (1981) considers the survival of the glowworm *Lampyris noctiluca* in Britain may be threatened by outdoor lighting and recommends establishing lamp-free reserves such as sheltered hollows as a protection measure.

There is also a considerable quantity of theoretical experimental data which suggests that photoperiodic responses are of great importance throughout all insect groups in regulating general activity, physiology and metabolism, mating behaviour, oviposition, development and life cycle (including effects of day length on post embryonic growth rate, moulting, number of generations and diapause) pigmentation and body form, polymorphism, species distribution etc. (Beck, 1980).

## CONCLUSIONS

Summary of the effects of Outdoor lighting on insects (after Frank, 1988)

- Outdoor lighting has been shown to disturb flight, navigation, vision, migration, oviposition, mating, feeding and crypsis in some moths. It may also disrupt circadian (daily) rhythms and photoperiodism.

**The responses of different species are varied so it is not possible to generalise as to the effects.**

- The attractiveness of lights varies for different insects and with the type of light source (See Table 1: Types of street lighting and their effects on insects).
- Isolated lights attract more individual insects than clusters of lights and high  background illumination may even repel some moths.

**The limitations of the evidence.**

- The problem is complex and it is not possible to generalise as to the effects on different species.
- The emphasis of much of the work done has been directed either towards control of pest species or enhanced food production.
- Manipulation of biological clocks seems to occur more readily in the laboratory than outdoors where temperature and other factors cannot be controlled.
- The problems may be underestimated due to apparent benefits to some species e.g. fast flying bat species feeding at light.
- There is a difficulty in making retrospective analyses of the effects on populations which may have already occurred.

**THE WAY FORWARD**

Major steps to reduce light pollution are now being taken in North America and in Europe, though the ecological problems of artificial lighting have only recently begun to be recognised. I first became interested in this topic in 1992 when I joined the staff at Hertfordshire Environmental Records Centre. There we became increasingly concerned at the proliferation in the number of applications for floodlighting/security lighting appearing in the planning lists of sponsoring District Councils. In 1994

Dacorum District Council hosted a symposium on Artificial Lighting and as a direct consequence of this I produced a paper on the subject for Hertfordshire Environmental Records Centre (Outen, 1995). Interest in this subject has grown rapidly since that time.

In Hertfordshire we have frequently challenged artificial lighting schemes where we considered they might be ecologically detrimental. One such instance was highlighted in the Department of the Environment Report *Lighting in the Countryside: Towards Good Practice (1997)*. In Europe, particularly Holland, Sweden and Denmark the problem is being taken very seriously. The Dutch Ministry of Transport and Water Resources commissioned a major desk top study on Road Lighting and Nature (Molenaar *et al.,* 1997) as a pilot project and is now funding two further phases involving field research.

In Britain the lighting industry is itself beginning to recognise the potential problem and issues some guidance and control recommendations (Institution of Lighting Engineers 1992; Department of Transport 1992; Department of the Environment 1997). The Institution of Lighting Engineers recently published a paper by me on the subject which has subsequently evoked many responsible and positive responses from the industry (Outen, 1997).We consider that Local Planning Authorities should at least be aware of ecological implications in considering future planning applications involving artificial lighting, though under current guidelines relating to Ecology (PPG9 etc.) it can be difficult to recommend outright refusal of a planning application on lighting grounds alone. We would however suggest that compromise agreements be sought where appropriate.

English Nature make the following recommendations with regard to the effect of lighting on insect populations and these guidelines are also beneficial with respect to most other groups:

- Query the need to install lighting near potentially vulnerable sites and oppose this.

- Use low-pressure sodium lamps with as low brightness as legally permissible.

- Fit shades to restrict light to where it is needed only. Astronomers also recommend this practice to counter light pollution.

- Fit ultra-violet filters to mercury lamps (sodium lamps emit negligibly in the UV) or suggest that these are changed to low-pressure sodium lamps.

- Turn off lamps close to vulnerable sites outside key periods of human activity if this does not put members of the public at risk.

English Nature suggest that so far as insects are concerned there is seldom justification for action to subdue lighting in areas of low wildlife interest and that the greatest problem is close to sites of high conservation value or to known populations of rare species. However there may be specific situations where in the interests of other organisms e.g. bats, otters etc. particular consideration may need to be given. In addition wider environmental issues may need to be taken into account. English Nature suggest that *"the effects of road lighting close to sites of high conservation interest should perhaps be considered when Conservation Officers are replying to consultations"* (Drake, 1994).

There remains a need for further field research into this problem and this would require funding. Should this be provided by Local or National Government, Public Charity or the Lighting Industry?

## ACKNOWLEDGEMENTS

Many individuals have been generous in their provision of information for this paper. Staff at Hertfordshire Environmental Records Centre have provided helpful comment and advice. Dacorum District Council provided the stimulus and encouragement for work on this subject. Margreet Westerhuis of Watling Chase Community Forest has kindly assisted with translation of parts of the major Dutch paper on the subject (Molenaar *et al.*, 1997).

## REFERENCES

**Agee HR. 1972.** Sensory response of the compound eye of adult *Helothis zea* and *H.virescens* to ultraviolet stimuli. *Anals of the Entomological Society of America.* **65:** 701-705.

**Aldrich JW, Graber RR, Munro DA.** *et al.* **1966.** Report of Committee on Bird Preservation *Auk* **83:** 465-467.

**Amies P. 1990.** Grey Herons hunting by artificial light. *British Birds* **83** (10): 425.

**Andresen JW. 1978.** Do streetlights turn city trees into late-growing insomniacs. *Amateur Forestry.* **84**:12.

**Axell HE. 1964.** *Night migrants at the old light, in Dungeness Bird Observatory* Dungeness Bird Observatory Committee.

**Baker J. 1990.** Toad aggregations under street lamps. *British Herpetological Society Bulletin.* **31**: 26-27.

**Baker RR, Sadovy Y. 1978.** The distance and nature of the light trap response of moths. *Nature.* **276:** 818-821.

**Balassina D. 1984.** *Amphibians of Europe: a colour field guide.* Devon: David & Charles.

**Beasley LJ, Zucker I. 1984.** Photoperiod influences the annual cycle of the male pallid bat (*Antrozous pallidus*). *Journal of Reproductive Fertility.* **70:** 567-573.

**Beasley LJ, Peltz KM, Zucker I. 1984.** Circannual rhythms of body weight in Pallid bats. *American Journal of Physiology.* **246:** R955-958.

**Beaty HH, Lilly JH, Calderwood DL. 1951.** Use of radiant energy for Corn Borer control. *Agricultural Engineering.* **32:** 421-422, 426, 429.

**Beck SD. 1980.** *Insect Photoperiodism* 2nd edn. London:Academic Press.

**Bernhard CG, Ottoson D. 1960.** Comparative studies on dark adaptation in the compound eyes of nocturnal and diurnal Lepidoptera. *Journal of General Physiology.* **44:** 195-203.

**Blest AD. 1963.** Longevity, palatability and natural selection in five species of New World saturnid moth. *Nature* **197:** 1183-1186.

**Bratton JH. 1990.** A review of the scarcer Ephemeroptera and Plecoptera of Great Britain. *Research and Survey in Nature Conservation.* NCC.

**Bretherton RF. 1951.** Our lost Butterflies and Moths *Entomological Gazette.* **2:** 211-240.

**Bretherton RF. 1954.** Moth traps and their lamps: an attempt at comparative analysis *Entomological Gazette.* **5:** 145-154.

**Brown C.H. 1976** A survey of the Sphingidae of Sanibel Island, Florida. *Journal of the Lepidoptera Society* **30:** 230-233.

**Campbell NA. 1990.** *Biology* (2nd edn).Benjamin Cummings Pub. Inc.

**Cantelo WW. 1974.** Blacklight traps as control agents: An appraisal. *Bulletin of the Entomological Society of America.* **20:** 279-282.

**Cathey HM, Campbell LE. 1975.** Effectiveness of five vision-lighting sources on photoregulation of 22 species of ornamental plants. *Journal of the American Society of Hortcultural Science.* **100:** 65-71.

**Clarke WE. 1912.** *Studies on Bird Migration* Vols 1 & 2. London: Gurney & Jackson.

**Collins D. 1934.** Iris-pigment migration and its relation to behaviour in the codling moth. *Journal of Experimental Zoology.* **69:** 165-167.

**Croft TA. 1978.** Night time Images of the earth from space. *Scientific American.* Jul: 86-98.

**Crowson RA. 1981.** *The Biology of Coleoptera.* New York: Academic Press.

**Cui G, Wardle CS, Glass CW** *et al.* **1991.** Light level thresholds for visual reaction of mackerel, *Scomber scombrus* L., to coloured monofilament nylon gillnet materials. *Fisheries Research* **10:** 255-263.

**Department of the Environment. 1997.** *Lighting in the Countryside: Towards Good Practice.* HMSO.

**Department of Transport. 1992.** *Reducing the environmental intrusion of road lighting (initial draft leaflet).* Department of Transport.

**Dick MH, Donaldson W. 1978.** Fishing vessel endangered by Crested Auklet landings. *Condor.* **80:** 235-236.

**Drake M. 1994.** Impact of Outdoor Lighting on Insect Populations. In: *Species Conservation Handbook.* Cambridge: English Nature.

**Egluchi E, Horikoshi T. 1984.** Comparison of stimulus-response (V-log I) functions in five types of Lepidoptera compound eyes (46 species). *Journal of Comparative Physiology.* **154:** 3-12.

**Elkins N. 1983.** *Weather & Bird Behaviour.* London: T & AD. Poyser.

**England MD. 1978.** Birds feeding by artificial light. *British Birds* **71:** 88.

**Farner DS. 1964.** The photoperiodic control of reproductive cycles in birds. *American Science.* **52:** 137-156.

**Fatzinger CW. 1979.** Circadian rhythmicity of sex pheromone release by *Dioryctria abietella* (Lepidoptera: Pyralidae (Phycitinae) and the effect of a diel light cycle on its precopulatory behaviour. *Annals of the Entomological Society of America.* **66:** 1147-1153.

**Feltwell J. 1996.** A hazard to moths on the Lozere massif *British Journal of Entomological Natural History.* **9:** 103-105.

**Fenton MB, Rautenbach IL. 1986.** A comparison of the roosting and foraging behaviour of three species of African insectivorous bats. *Canadian Journal of Zoology.* **64:** 2860-2876.

**Ford EB. 1972.** *Moths* 3rd Edn. London: Collins.

**Frank KD. 1988.** The Impact of Lighting on Moths: An Assessment. *Journal of the Lepidopterist's Society.* **42.** 63-93.

**Freeman HJ. 1981.** Alpine Swifts feeding by artificial light at night. *British Birds* **74:** 149.

**Friedrich E. 1986.** *Breeding Butterflies and Moths.* Harley Books.

**Geier PW. 1960.** Physiologic age of codling moth females (*Cydonia pomonella*) caught in bait and light traps. *Nature.* **185** (4714): 709.

**Glass CW, Wardle CS, Mojsiewicz WR. 1986.** A light intensity threshold for schooling in the Atlantic mackerel, *Scomber scombrus. Journal of Fish Biology.* **29** (supp A): 71-81.

**Glass CW, Wardle CS. 1989.** Comparison of the Reactions of Fish to a Trawl Gear, at High and Low Light Intensities. *Fisheries Research* **7:** 249-266.

**Graham HM, Glick PA, Martin DF. 1964.** Nocturnal activity of adults of six lepidopterous pests of cotton as indicated by light trap collections. *Annals of the Entomological Society of America.* **57:** 328-332.

**Green D. 1978.** Robins feeding young at night. *British Birds.* **71:** 83-84.

**Hazard T, Eddy R. 1950.** Modification of the sexual cycle in the brook trout (*Salvelinus fontinalis*) by control of light. *Transactions of the American Fisheries Society.* **80:** 158-162.

**Heath J. 1974.** A century of change in the Lepidoptera. In: Hawksworth L. ed. *The changing flora and fauna of Britain.* London: Academic Press.

**Heinig S. 1982.** Uberwintert *Macroglossum stellatarum* auch als puppe? *Ent.Z; Frankff. a.M.* **88:** 53-62.

**Hendry A. 1984.** Light Pollution: A status report. *Sky and Telescope* **June:** 504-507

**Herms WB. 1932.** Deterrent effect of Artificial light on the codling moth. *Hilgardia.* **7:** 263-280.

**Hienton TE. 1974.** *Summary of investigations of electric insect traps.* U.S.Dept. of Agrculture Technical Bulletin. 1498.

**Hill D. 1992.** *The impact of noise and artificial light on waterfowl behaviour: a review and synthesis of available literature.* British Trust for Ornithology.

**Hosny MM. 1959.** Review of results and a complete list of Macrolepidoptera caught in two ultra-violet light traps during 24 months at Rothamsted, Hertfordshire. *Entomologist's Monthly Magazine.* **95:** 226-237.

**Howe WH. 1959.** A swarm of noctuid moths in south-eastern Kansas. *J.Lepid.Soc.* **13:** 26.

**Howell JC, Laskey AR, Tanner JT. 1954.** Bird Mortality at airport ceilometers *Wilson Bulletin.* **66:** 207-215.

**Hsiao HS. 1972.** *Attraction of Moths to light and Infra-red radiation.* San Francisco: San Francisco Press.

**Institution of Lighting Engineers. 1992.** *Guidance notes for the reduction of light pollution.* Institution of Lighting Engineers.

**Jaeger RG, Hailman JP. 1973.** Effects of intensity on the phototactic response adult anuran amphibians: a comparative survey. *Zeitschrift Tierpsychologie* **33:** 352-407.

**James P. 1956.** Destruction of Warblers on Padre Island, Texas, in May 1951. *Willson Bulletin.* **68:** 224-227.

**Knox A. 1990.** Extended foraging period of nesting swallows. *British Birds* **83(4):** 166.

**Lack D. 1965.** *The life of the Robin* 4th edn. London: Collins.

**Langmaid JR. 1959.** Moths of a Portsmouth garden - a four year appreciation. *Entomological Gazette.* **10:** 159-164.

**Laskey AR. 1956.** Bird Casualties at Smyrna and Nashville ceilometers, 1955. *Migrant.* **27:** 9-10.

**Lofts, Merton. 1968.** Photoperiodic and physiological adaptions regulating avian breeding cycles and their ecological significance. *Journal of Zoological Society of London.* **155:** 327-94.

**Lukefahr M. & Griffin J. 1957.** Mating and oviposition habits of the pink bollworm moth. *Journal of Economic Entomology.* **50:** 487-490.

**McFarlane RW. 1963.** Disorientation of loggerhead hatchlings by artificial road lighting. *American Society of Ichthyologists and Herpetologists.* **Vol 1963:** 153.

**McGeachie WJ. 1988.** A remote sensing method for the estimation of light-trap efficiency. *Bulletin of Entomological Research.* **78:** 379-385.

**Mead CM. 1983.** *Bird Migration.* London: Country Life Books.

**Molenaar JG, Jonkers DA, Henkens RJH. 1997.** *Wegverlichting en Natuur* DWW Ontsnipperingsreeks. 34.

**Morrison DW. 1978.** Lunar phobia in a neotropical fruit bat *Artibeus jamacensis* (Chiroptera: Phyllostomidae). *Animal Behaviour.* **26:** 852-855.

**Nau BS, Boon CR, Knowles J. eds. 1987.** *Bedfordshire Wildlife.* Castlemead: Ware.

**Nemek SJ. 1969.** Use of artificial lighting to reduce *Heliothis* spp. populations in cotton fields. *Journal of Economic Entomology.* **62:** 1138-1140.

**Nemek SJ. 1971.** Effects of lunar phases on light-trap collections and populations of bollworm moths. *Journal of Economic Entomology.* **64:** 860-86.

**Nomura K. 1969.** Studies on orchard illumination against fruit-piercing moths *Review of Plantation Protection Research. (Tokyo)* **2:** 122-12.

**Odum EP. 1971.** *Ecology.* Holt International.

**Outen AR. 1995.** *The Possible Ecological Effects of Artificial Lighting.* Hertfordshire Environmental Records Centre.

**Outen AR. 1997.** The Ecological Impact of Artificial Light. *Lighting Journal.* **62 (5):** 11-15.

**Pettingill OS. 1970.** *Ornithology in Laboratory and Field.* Minneapolis: Burgess Publishing Company.

**Pyle RM, Bentzien M, Opler P. 1981.** Insect Conservation. *Annual Review of Entomology.* **26:** 233-258.

**Rawson HE. 1923.** A birds song in relation to light. *Transactions of the Hertfordshire Natural History Society.* **17:** 363-365.

**Reed JF. 1981.** Effectiveness of highway lighting in reducing deer-vehicle accidents. *Journal of Wildife Management.* **45(3):** 721-726.

**Reed JR, Hailman JP, Sincock JL. 1985.** Light attraction in procellariform birds: reduction by shielding upward radiation. *Auk.* **102:** 377-383.

**Rees EC. 1982.** The effect of photoperiod on the timing of spring migration in the Bewicks Swan. *Wildfowl.* **33:** 119-132.

**Reid K. 1969.** *Natures Network.* Aldous.

**Reinhold JO. 1993.** Lantaarnpalen en laatvliegers. *Nieuwsbrief VLN.* **15(5):** 2-5.

**Robinson HS, Robinson PJM. 1950.** Some notes on the observed behaviour of Lepidoptera in flight in the vicinity of light sources together with a description of a light trap designed to take entomological samples. *Entomological Gazette.* **1:** 3-20.

**Robinson PJM. 1952.** On the behaviour of night-flying insects in the neighbourhood of a bright source of light. *Proceediongs of the Royal Entomological Society of London.* **A27:** 13-21.

**Rydell J. 1991.** Seasonal use of illuminated areas by foraging bats *Eptesicus nilsonii. Holarctic Ecology.* **14(3):** 203-207.

**Rydell J, Racey PA. 1993.** Street Lamps and the feeding ecology of Insectivorous Bats. In: *Recent Advances in Bat Biology. London:* Zoological Society of London (Symposium Abstracts).

**Schachinger F. 1962.** Geel autolicht tegen wildverliezen op de weg. *De Nederlandse Jager.* **67(1):** 19.

**Scott DK. 1980.** The behaviour of Bewick's Swans at the Welney Wildfowl Refuge, Norfolk and on the surrounding fens: a comparison. *Wildfowl.* **31:** 5-18.

**Shropshire W. 1977.** Photomorphogenesis. In: Smith KC. ed. *The Science of photobiology.* New York: Plenum. 281-312.

**Sinnadurai S. 1981.** High pressure sodium street lights affect crops in Ghana. *World Crops.* **(Nov/Dec):** 120-122.

**Skinner B. 1984.** *Colour Identification Guide to Moths of the British Isles.* London: Viking.

**Skopik SD, Takeda M. 1980.** Circadian control of oviposition activity in *Ostrinia nubilalis. American Journal of Physiology.* **239:** R259-R264.

**Speakman JR. 1991.** Why do insectivorous bats in Britain (UK) not fly in daylight more frequently? *Functional Ecology.* **5(4):** 518-524.

**Stewart PA, Lam JJ, Blyth JI. 1969.** Influence of distance on attraction of Tobacco Hornworm and Corn Earworm moths to radiations of a black-light lamp. *Journal of Economic Entomology.* **62:** 58-61.

**Swift SM. 1980.** Activity patterns of pipistrelle bats (*Pipistrellus pipistrellus*) in north-east Scotland. *Journal of the Zoological Society of London.* **190:** 285.

**Taylor LR, Carter CI. 1961.** The analysis of numbers and distribution in an aerial population of Macrolepidoptera. *Transactions of the Royal Entomological Society of London.* **113:** 369-386.

**Taylor LR, French RA, Woiwood IP. 1978.** The Rothamsted insect survey and the urbanisation of land in Great Britain. In: Frankie GW, Koehler CS. eds. *Perspectives in Urban Entomology.* New York: Academic Press. 31-56.

**Telfer TC, Sincock JL, Vernon Byrd G, Reed JR. 1987.** Attraction of Hawaiian seabirds to light: conservation efforts and effects of moon phase. *Wildlife Society Bulletin.* **15:** 406-413.

**Terres JK. 1956.** Death in the Night. *Audubon.* **58:** 18-20.

**Turrill WB. 1948.** *British Plant Life.* London: Collins.

**Tuttle JP. 1985.** Maintaining species integrity between sympatric populations of *Hyalophora cecropia* and *Hyalophora columbria* (Saturniidae) in Central Michigan. *Journal of the Lepidopterist's Society.* **39:** 65-84.

**Vaisanen R, Hublin C. 1983.** The effect of continuous light trapping on moth populations. A mark-recapture experiment on *Hydraecia petasites* (Lepidoptera: Noctuidae). *Notulae Entomology.* **63:** 187-191.

**Verheijen FJ. 1980.** The moon: a neglected factor in studies on collisions of nocturnal migrant birds with tall lighted structures and with aircraft. *Die Vogelwarte.* **30:** 305-320.

**Verheijen FJ. 1981.** Bird kills at tall lighted structures in the USA in the period 1935-1973 and kills at a Dutch lighthouse in the period 1924-1928, show similar lunar periodicity. *Ardea.* **69:** 199-203.

**Verheijen FJ. 1985.** Photopollution: Artificial light optic control system fail to cope with. Incidents, causationss, remedies. *Experimental Biology.* **44:** 1-18.

**Waring P. 1994.** Moth traps and their use. *British Wildlife.* **5(3):** 137-148.

**Webb NR, Malt DC. 1977.** Five years light-trapping of moths at Furzebrook. In: *Institute of Terrestrial Ecology Annual Report.* NERC.

**Williams CB. 1939.** An analysis of four years captures of insects in a light trap. *Transactions of the Royal Entomological Society of London.* **86:** 79-132.

**Williams CB, Singh BP, El Ziady S. 1956.** An investigation into the possible effects of moonlight on the activity of insects in the field. *Proceedings of the Royal Entomolgical Society of London.* **31:** 135-144.

**Woiwood IP, Riley AM. 1992.** The Rothamsted Insect survey light trap network. in Harding AT. ed. *Biological Recording of changes in British Wildlife.* (ITE symposium 26). HMSO.

**Wolfe JL, Summerlin CT. 1989.** The influence of lunar light on nocturnal activity of the Oldfield Mouse. *Animal Behaviour.* **37:** 410-414.

**Worth CB, Muller J. 1979.** Captures of large moths by an Ultra-violet light trap. *Journal of the Lepidopterist's Society.* **33:** 261-265.

# 14

# Design, build, fund and operate: ecological issues during the construction of privately financed road schemes

## J WYNN & W LATIMER

WS Atkins, as part of the 'Connect' consortium, have recently been involved in three major 'Design, Build, Fund and Operate' (DBFO) schemes: the A50 Doveridge By-pass; the A35 Tolpuddle-Puddletown By-pass and the A30 Honiton-Exeter Improvement Scheme. As part of the design team, WS Atkins Environment provided ecologists to supervise and carry out the ecological requirements of the contract. This involved:

- Survey for protected species prior to site clearance.

- Advising on mitigation in the event that protected species are found.

- Advising design and construction engineers on methods and timing of operations to minimise ecological disturbance.

- Liaison with the Environment Agency and English Nature to establish methods for dealing with watercourses and protected species.

- Implementing mitigation measures such as habitat creation, barn owl boxes and bat boxes.

- Ecological monitoring, throughout the construction period and, in some cases beyond, of mitigation schemes such as bat boxes, barn owl boxes and badger fencing.

## BENEFITS

There are several ecological benefits attached to this method of financing road construction, although there are many ethical and political questions attached to the practice of funding road building through private finance. These ecological benefits include:

- The requirement for close supervision and control of ecological aspects. Failure to comply with the requirements of the contract can result in the construction company accumulating penalties, which may adversely affect their chances of winning subsequent contracts.

- Ecological assessment and survey can be on going throughout the life of the project, particularly with regard to protected species.

- Active measures recommended in the Environmental Statement can be pursued.

- As the footprint of a road development is often less discrete than for other developments, such as business parks and housing developments, the presence of an ecologist on site provides the opportunity to assess changes in the final engineering design, which necessitate extra or different land take. These may include accommodation roads, access tracks and temporary structures such as construction compounds and batching plants, which are often not included in the original Environmental Statement.

The company also addressed the long-term perspectives of road building in this country. Key issues were identified as:

- The review of the road building budget to investigate the best methods of developing an integrated transport policy to achieve the 8% reduction in greenhouse gases agreed at the Kyoto UN Framework Convention on Climate Change.

- Recognition by the Government that cars will remain an important element of the UK's transport system. This will, inevitably, mean the construction of new roads and the upgrading of existing roads. Both of these activities have ecological implications. It is essential, therefore, that the formulation of guidelines for the protection and mitigation of protected and endangered species and valuable habitats continues. The experience and practice developed through the DBFO approach can form a useful basis for the formulation of these guidelines.

- The Environmental Impact Assessment and the Environmental Statement set the baseline for this. It is therefore essential that the impact assessment addresses all ecological issues and provides for best standards of ecological mitigation. At present the EIA rarely covers the complete land take.

- Additionally, although the extent, location and activity of protected species are documented in the Environmental Statement, there is often a long period of time between completion of the statement and the start of construction. In the intervening period, it is not unusual for the ecological situation to have changed. This can be easily addressed at the start of construction on a DBFO scheme, as any change can be identified during the on-going surveys, which are required by the DBFO contract.

## CONCLUSION

We hope these schemes have shown that progress has been made in the area of environmental protection through DBFO schemes. There is, however, still room for improvement in the environmental assessment of all contingent effects. This must take place through both the design and construction phases of road building if it is to be fully effective.

# 15

# Environmental impacts of transport infrastructure: habitat fragmentation and edge effects.

## P G ANGOLD

Roads may act as habitats, as linear corridors or as barriers to the dispersal of animal and plant species (e.g. Vermeulen, 1995). There are many techniques for aiding the movement of vertebrates across roads, and some for invertebrates (e.g. Bekker *et al.*, 1995). Less is known about the movement of plant propagules, although the spread of some species along road verges, in particular the increasing distribution of coastal species along road verges, is widely reported. There is now increasing public concern over the environmental impacts of roads and road construction in the UK. This paper quantifies the edge effects generated by a road by investigations into pollutant dispersal, vegetation structure and composition, the growth of selected plant species, and the soil nutrient status. The results show that unless roads are designed and built so as to buffer the impacts of ongoing vehicular use, adverse effects may be seen to considerable distances beyond the verge in delicate or vulnerable ecosystems.

## LOWLAND DRY HEATH

British lowland dry heath (a semi-natural ecological community dominated by dwarf shrubs, e.g. *Calluna vulgaris*) was chosen for study because it is of international importance for conservation, and has suffered extensive fragmentation (Webb & Haskins, 1980), often by roads. It is also a naturally nutrient poor and relatively unproductive community, which will be particularly sensitive to disturbance or to eutrophication. The impact of roads and normal road use in adjacent heathland was investigated in detail by studies of individual species and ecosystem function at different distances from the road.

The environmental impact of a road is complex, beginning with the physical impact of habitat loss, and associated changes to soil structure, groundwater flow and soil-water relations, which occur during construction. After construction of the road, there is an ongoing impact of pollutants from motorised vehicles. Vehicle exhaust emissions include solid particle fuel (particularly from diesel fuel), carbon monoxide, carbon dioxide, sulphur dioxide, oxides of nitrogen, organic gases such as ethylene, and heavy metals such as lead. Other pollutants include ozone from photochemical reactions, cadmium from vehicle tyres, and phenols and bacteria from the road surface (Watkins 1991). Given the wide range of pollutants and potential pollutants from roads, and the possibility for secondary and synergistic effects, very few generalisations can be made concerning the effects of pollutants on biota near roads. Each species will respond to different pollutants in different ways, and even different individuals of the same species may respond differently according to stages in life history or differences in local environmental factors (Wellburn, 1990; Rowland *et al*, 1985). The environmental impact of a road on adjacent heathland has therefore been studied in some detail. Investigations have been undertaken into pollutant dispersal, vegetation structure and composition, the growth of selected plant species, and the soil nutrient status to gain an understanding of the overall impact of the road at an ecosystem level.

## Method

The fieldwork was undertaken in an extensive area of heathland on either side of the A31, the major trunk road through the New Forest in Hampshire, UK. This road carried 34,661 vehicles during a measured 12 hour flow in October 1990. Five research sites were established adjacent to this road, each with eight replicate sampling stations at 1.5, 5, 10, 25, 45, 80, 150 and 200 metres from the road.

The pollution from vehicle exhausts was assessed by the use of nitrogen diffusion tubes to measure the flux of a secondary pollutant of vehicle exhausts: nitrogen dioxide. The tubes contained grids impregnated with triethanolamine/acetone solution to absorb nitrogen dioxide, and a tube was left in position at each sampling station for 3 weeks. Nitrogen dioxide was then extracted using sulfanilamide and N-1-naphthylethylene-diamine-dihydrocholride reagents, and the flux of nitrogen dioxide was measured colorimetrically (Robbins 1985).

The plant species composition at each sampling station was recorded within a 50cm x 50cm quadrat, using the Domin scale of cover abundance, and analysed using the multivariate analysis package DECORANA (Hill, 1979). The extent of the edge effect from the road into the heath was assessed by plots of DECORANA score against distance from the road.

The performance of selected species (*Calluna vulgaris, Molinia caerulea*, and *Cladonia impexa*) was assessed. The annual growth increment of *Calluna* and *Molinia* was recorded, and the nitrogen and phosphorus content of the shoots was assessed by autoanalysis following a sulphuric acid:hydrogen peroxide micro-Kjehldahl procedure, using phenate and sodium hypochloride reagents for the analysis of nitrogen, and molybdate and vanadate reagents for the analysis of phosphorus (Allen *et al*, 1974). The diameter of the clump was recorded for *Cladonia* samples.

Soil nutrient levels were also investigated at each sampling site: carbon content was assessed by loss on ignition, and nitrogen and phosphorus by autoanalysis of the micro-Kjehldahl acid digest (procedure as for vegetation samples).

The implications of edge effects were investigated using the maps of heathland habitat present at various times in the Poole Basin (Webb & Haskins, 1980). The area and perimeter of each heathland fragment extant in 1759, 1811, 1934, 1960 and 1978 were measured and the core area model developed by Laurance & Yensen (1991) used to calculate the core area of each fragment after subtracting the edge habitat for theoretical edge effects of 0, 25, 50, 100 and 200 metres.

## Results

The atmospheric flux of nitrogen dioxide was strongly elevated near the road, to 60ppb/hr, and decreased exponentially with distance from the road at each site (Figure 1).

The cover abundance of *Calluna vulgaris* and of *Cladonia impexa*, and the mean number of species of lichen per quadrat, decreased near the road, whereas the cover abundance of *Molinia caerulea* increased (Figure 2). *Calluna* plants near the road showed greater annual growth increments and nitrogen concentration (ANOVA $P<0.001$; mean results 61mm, 1375mg /100g nitrogen at 25m; and 41mm, 1295mg/100g at 80m from the road). *Molinia* plants also tended to be larger near the road (ANOVA $P<0.05$, mean results 406mm and 352mm), but the tissue nutrient concentration was not significantly different. Clumps of the lichen *Cladonia impexa* were fewer and smaller near the road, individual stems were significantly shorter with fewer branches (ANOVA $P<0.01$).

The soil carbon content decreased near the road (ANOVA $P<0.001$, mean values 29% and 40%). The carbon:nitrogen ratio also decreased (ANOVA $P<0.001$, mean values 24.7 and 28.0), as did the carbon: phosphorus ratio (ANOVA $P<0.001$; mean values 86 and 118). The changes in the heathland community were detectable up to 200m away from the road (e.g. Figure 2).

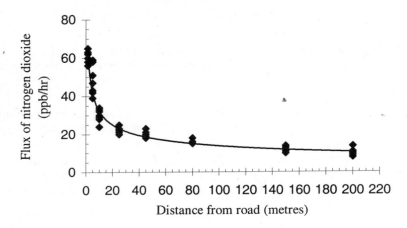

**Figure 1.** Change in flux of atmospheric nitrogen dioxide with distance from the road

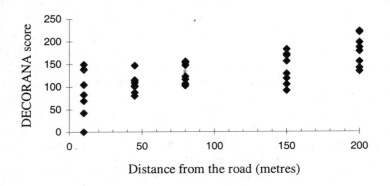

**Figure 2.** Change in plant species composition of heathland. The y axis is the detrended correspondence analysis (DECORANA) quadrat score. (Regression data: $r^2=0.37$; P<0.001) The quadrat score is the weighted mean species score. Low scoring species: the grasses *Molinia caerulea* and *Festuca rubra*. High scoring species: *Calluna vulgaris*, mosses, lichens.

The rate of loss of core area increases with increasing habitat fragmentation (Figure 3). The proportion of core area is shown to be drastically reduced by habitat fragmentation; the 7 large fragments in 1759 would have been 82% core area, but following further fragmentation to 435 fragments, only 10% of the reduced amount of extant heath could be core habitat.

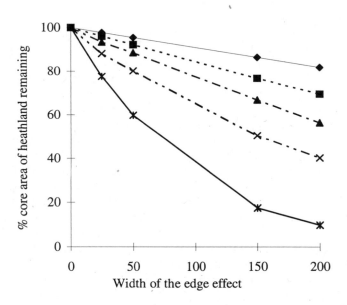

**Figure 3.** The percent core area of heathland remaining in the Poole Basin, U.K. over time and assuming a range of edge effects. Key: ◆=1759, 7 fragments; ■=1811, 13 fragments; ▲ = 1934, 17 fragments; x =1960, 97 fragments; * =1978, 435 fragments

## DISCUSSION

There were changes in plant species composition, plant performance and soil nutrient levels near the road. The grass species *Molinia caerulea* increased in abundance and vigour, whereas other plants such as the definitive heathland species *Calluna vulgaris* decreased in abundance. The vascular plants generally showed increased growth rates; in contrast the lichen species *Cladonia impexa* grew less luxuriantly near the road. There was also a decline in the abundance and diversity of the overall lichen flora near the road.

The organic content of the soil decreased near the road, which in view of the increased biomass produced by enhanced plant growth suggests an increased rate of decomposition and nutrient cycling. The soils nearer the road showed greater carbon:nitrogen and carbon:phosphorus ratios, and a decrease in organic content, indicating an increase in the availability of nutrients near the road. The primary edge effect generated by the road was therefore one of eutrophication.

Oxides of nitrogen, which reached a flux of 60ppb/hr immediately adjacent to the road, were shown to exceed the winter critical level of 20ppb/hr, which may cause damage to vegetation (Anon, 1990); and many species of lichen are known to be particularly vulnerable to atmospheric pollution. In addition, with increased nutrient availability, the more competitive plant species such as *Molinia* will generally become dominant through a greater increase in relative productivity (Aerts & Caluwe, 1989). Thus a positive feedback is established near roads, in which nitrogen from vehicle exhausts causes increases in the growth rate of *Molinia,* resulting in a higher rate of decomposition (Aerts, 1989; Bobbink *et al.*, 1990), and greater nitrogen availability, which further increases the rate of growth of *Molinia*. This results in an edge effect, a gradient of change in the heathland vegetation extending from the road at least 200 metres into the heathland habitat to either side of the road. A study (Angold, 1997) of other New Forest roads has shown a positive log-linear relationship between the extent of the edge effect and the traffic volume, and therefore a direct relationship between  the extent of the edge effect and the amount of atmospheric pollution from vehicle exhausts.

The investigation into the impact of this scale of edge effect upon a fragmented network of habitat patches showed that the environmental impact is very severe with increasing fragmentation and isolation of habitat patches (Figure 3).  In a very fragmented modern landscape of the Poole Basin, up to 90% of the area regarded as heathland may thus be disturbed edge habitat, and even of the remaining 10%, the core area in some fragments may be too small to support the full heathland community (Webb & Vermaat, 1990). Thus building or widening of roads can have severe environmental impacts through habitat fragmentation and degradation at the edges of the remaining habitat by increasingly severe edge effects as well as by direct habitat loss. Buffer zones will be needed to absorb the edge effect near important sites if biodiversity is to be preserved in a landscape with extensive linear transport infrastructure.  Further research is needed to optimise the use of such buffer zones and minimise the extent of edge effects by designing buffer zones to absorb the environmental impacts of roads with maximum efficiency.

## ACKNOWLEDGEMENTS

Credits to Conder Conservation Trust for funding, the U.K. Forestry Commission for fieldwork access, PJ Edwards and IF Spellerberg for long-term help and encouragement. Thanks to 'IENE' (Infra Eco Network Europe) for stimulation and debate on the environmental impact of roads at an international level.

## REFERENCES

**Aerts R. 1989.** The effect of increased nutrient availability on leaf turnover and aboveground productivity of two evergreen ericaceous shrubs. *Oecologia* **78**: 115-120.

**Aerts R, de Caluwe H. 1989.** Above-ground productivity and nutrient turnover in of *Molinia caerulea* along an experimental gradient of nutrient availability. *Oikos* **57**:310-318.

**Allen SE, Grimshaw HM, Parkinson JA, Quarnby C. 1974.** *Chemical analysis of ecological materials.* Oxford: Blackwell Scientific Publications.

**Angold PG. 1997.** The impact of a road upon adjacent heathlands vegetation: effects in plant species composition. *Journal of Applied Ecology* **34:** 409-417.

**Anon. 1990.** *Oxides of nitrogen in the United Kingdom.* London: Department of the Environment.

**Bekker H, Hengel VDB, Bohemen VH, ven der Sluijs H. 1995.** *Nature across motorways.* Public Works & Water Management. Delft: Netherlands.

**Bobbink R, Heil G, Raessen M. 1990.** *Atmospheric deposition and canopy exchange in heathland ecosystems.* Utrecht: Elinkwijk BV.

**Hill MO. 1979.** *DECORANA - a Fortran program for detrended correspondence analysis and reciprocal averaging.* Ecology and Systematics, Ithaca, New York: Cornell University.

**Laurance WF, Yensen E. 1991.** Predicting the impacts of edge effects in fragmented habitats. *Biological Conservation.* **55:** 77-92.

**Robbins H. 1985.** Effects of roadside pollutants on disease/plant interactions. [PhD Thesis] Newcastle: University of Newcastle-upon-Tyne.

**Rowland A, Murray AJS, Wellburn AR. 1985.** Oxides of nitrogen and their impact on vegetation. *Reviews on Environmental Health.* **5**: 295-342.

**Vermeulen HJW. 1995.** Roadside verges. [Ph.D. Thesis] Landbouw: Netherlands.

**Watkins LH. 1991.** *Air pollution from road vehicles.* London: HMSO.

**Webb NR, Haskins LE. 1980.** An ecological survey of heathlands in the Poole Basin, Dorset, England in 1978. *Biological Conservation.* **17:** 153-167.

**Webb NR, Vermaat AH. 1990.** Changes in vegetational diversity on remnant heathland fragments. *Biological Conservation.* **51:** 253-264.

**Wellburn AR. 1990.** Why are atmospheric oxides of nitrogen usually phytotoxic and not alternative fertilisers? *New Phytologist.* **115:** 395-429.

# 16

# Roads as barriers

## P ANDERSON

Roads may change the direction of dispersal flow between populations, or decrease the dispersal flow across the landscape (Opdam, 1997). Species movements may be inhibited between parts of their habitats by the barrier effect of roads, for example, for amphibians, badgers or deer. The extent and significance of this barrier effect needs to be put into a broader context to identify its significance. The habitat patches (small woods, heaths, ponds, meadows etc.) which, in a cultural landscape, are scattered through the matrix of more intensively managed agricultural fields or urban areas, are already separated by land which is an inhospitable environment for many species. This ecological landscape already has a degree of resistance to the movement of plants and animals which will vary between species. Roads add a linear barrier to this resistance which is only sometimes an absolute barrier (Opdam, 1997). If the road is fairly permeable and the total resistance of the ecological landscape already high, the significance of the linear barrier may not be great. On the other hand, if the resistance of the landscape is low and there is free movement of species, then adding a major road structure could have significant effects for some species as, for example, when a road divides a significant habitat patch into two. The barrier effect is part of the wider issue of habitat fragmentation (Canters & Cuperus, 1997) and contributes to a reduction in quality rather than absolute destruction of habitats and assemblages.

## MOBILITY OF SPECIES

The significance of the barrier presented by a road relates to the relative mobility and behavioural needs of species. Species mobility can be regarded as a continuation of variation from the more mobile to the most sedentary. Species may vary in their mobility during their lifecycle and may be more mobile in a dispersal phase (as in many plants and invertebrates), or as part of a breeding or hibernation cycle (such as amphibians and some reptiles). In animals, mobility will also be relative to the size and hence extent of home range or territory. The larger this is, the more likely it is to overlap with areas of resistance in the landscape, including roads. Examples at one end of the mobility spectrum include a wide range of invertebrates, some birds and plants with seeds which are wind dispersed and for which the ecological landscape structure provides little hindrance to such dispersal and, on the other hand, to species which are confined to a few square metres and do not disperse far from this, such as some ancient woodland plants and sand lizards *Lacerta agilis*. The more mobile species would generally be unaffected by the barrier effects of roads. Paradoxically, roads probably do not form barriers to the least mobile species either, unless the road were to bisect their habitat patch since they cannot move far nor cross alien habitat types. These species exist essentially in closed populations. There is, however, a substantial core of species which has some mobility or dispersal capabilities, and which move between and throughout habitats, thus maintaining genetic diversity and avoiding inbreeding.

Some of these species have been found to exist in meta-populations in a fragmented landscape. Roads may create major barriers for some of these as they move across the environment which could contribute to the loss of sub-populations. If this occurs in several places within a short time period, local extinctions could occur.

The degree to which roads will act as barriers will depend on the width and amount of traffic, and indirectly, the amount of light and noise. The degree to which different species are affected by these features will determine the significance of the effect. A final aspect of mobility and roads which requires consideration is the group of species, mostly mammals, reptiles and amphibians, for which a road is not a physical barrier – they could cross easily, but where the amount of traffic results in a high casualty rate. In severe cases a local population can be completely lost, as in some amphibian and badger *Meles meles*, populations.

## THE EVIDENCE FOR A BARRIER EFFECT

Schonewald-Cox & Beucher (1992) consider roads to be a significant barrier to the dispersal of many plants and animals. Harris & Silva-Lopez (1992) suggest that a road carrying a high level of traffic will constitute a more serious barrier than a little used one. Most of the research on the barrier effect of roads has concentrated on mammals, although some data are available on reptiles, amphibians and invertebrates. Some effects can be extrapolated from other work, particularly on plants and birds.

## Mammals

Roads present a barrier to a wide range of animals, altering behaviour patterns and preventing dispersal. In the European context, Mader (1984) recounts how rarely yellow necked mice *Sylvaemus flavicolis*, and bank voles *Clethrionomys glareolus*, crossed a variety of roads, from a busy highway (250 vehicles/hour) down to a little used forest road. All the roads studied were situated in forests, and a mark and recapture method was used to determine movement patterns. None of the 121 marked and 35 recaptured animals crossed the 6m wide highway whilst, even on a non-public forest road, only 2 crossings were observed compared with 34 movements between trap rows. In a uniform 'control' area without a road, 28 animals passed over an imaginary road line. A similar barrier effect of a 5m road through a forest in Poland was found for bank voles (quoted by Merriam *et al.*, 1989) but here, yellow necked mice were not affected.

Oxley *et al.* (1974) found similar barrier effects in a study in Ottawa, Canada. Small forest mammals were reluctant to venture onto road surfaces where the distance between the forest margins at either side was greater than 20m, irrespective of the level of traffic. Wider roads were crossed nearly exclusively by medium sized animals. Swihart & Slade (1984) demonstrated that even a seldom used dirt track presented a major barrier to prairie voles *Microtus ochrogaster*, but that crossings of this track made by cotton rats *Sigmodon hispidus*, were density dependent. Although other American studies have confirmed the type of findings that Oxley *et al.* (1974) reported, and in habitats other than woodlands, such as deserts, this study documents the smallest road clearance (less than 3m) which inhibits some small mammal movements.

Although crossings by the cotton rats were much lower than expected, sufficient movement of animals occurred to maintain the gene flow in the habitats on either side of the track. These types of findings suggest that further research into some of Britain's small mammals is needed to estimate the extent and significance of barrier effects by our dense road network.

## Invertebrates

Very little work has been undertaken on the effects of roads, or any other transport corridor, on various invertebrates. Mader (1984) studied highly mobile forest carabids, all of which had the capacity to move the equivalent distance of the road width on foot. Over a 5 year period, 10,186 carabid beetles were marked and released but between none and very few individuals of different species ventured across the roads.

Dennis (1986) found a similar pattern in a study of the orange tip butterfly *Anthocharis cardomines*, trying to cross the M56 as it bridges the Bollin Valley on the Manchester/Cheshire boundary. Using a mark and recapture technique, he concluded that only about 2% of individuals approaching the motorway actually managed to cross it. Most specimens seen to climb high enough to cross the motorway were knocked back by the buffeting effect of the heavy traffic use (28 vehicles/minute each way at the time). However, Dennis considered that, possibly, one butterfly every 2 hours crossed the road and that this leakage is adequate for the transfer of genes.

The orange tips in Dennis's study were regarded as part of a mostly open population, but with some signs of the behaviour of a closed one. Munguira & Thomas (1992) found, in a study of butterflies and burnet moths on 12 A-road verges in Dorset and Hampshire, that 21 species could cross these roads, but that 9 species failed to do so. Of the more mobile species, between 10 and 32% of all recaptured adults had crossed the road, whilst 2 individuals re-crossed a road. All those species which crossed more frequently than expected are regarded as species of open, mobile populations, whilst those that crossed infrequently or not at all live mostly in closed populations.

In a study of the land snail, *Arianta arbustorum*, Baur & Baur (1990) found that movements were largely confined to the edges of two types of road - a paved, 8m wide, low traffic volume site, and an unpaved, 3m wide, track. Several individuals moved far enough to have crossed the roads, but only 1 recaptured individual crossed the paved road, and 2 the track (809 released and 29% recaptured altogether). In contrast, 41.7% of recaptures of the same species had crossed an overgrown, 0.3m wide, path only occasionally used by walkers. The conclusion was that the paved road, in particular, acted as a barrier to dispersal.

In a further study on ground beetles, this time including some wolf spiders *Lycosids* spp, Mader *et al.* (1990) investigated the barrier effect of narrow, paved, gravel or grassy agricultural roads, and a single-track railway used by 4 trains/hour in daytime only. The results indicate that all these barriers stimulate a lengthways dispersal and inhibit movements across them. The percentage of recaptures which had crossed varied from 3% to 17% in different ground beetle species, with *Pterostichus madidus* being the most inhibited. In the controlled release of the Lycosid, *Pardosa amentata*, none crossed the field track, although Lycosids can disperse more widely to a variable degree as juveniles by ballooning.

Mader *et al.* (1990) concludes that the rate of invasion of the species tested could be significantly diminished by roads acting as barriers. For those animals adapted to disperse, for whatever reason, the network of barriers tend to guide them parallel with roads or railway tracks and could reduce the average distance they can move if their energy supply becomes exhausted. As the distances between habitats increase, due to degradation or destruction, the chances decrease for the dispersal of many species.

## Other Species

The direct barrier effect of roads has not been examined for many other species, but some information is available. Corbett & Tamarind (1979) comment that sand lizards cannot recolonise fragmented sites due to the alien habitat in between, and roads are inferred as part of this barrier. Smooth snakes *Coronella austriaca*, will not cross main, busy roads either, but Ian Spellerberg (*pers. comm.*) considers that they probably do cross minor roads. Spellerberg & Phelps (1977) though, identify forest tracks as an apparent barrier to significant movement in one study site in Southern England, yet found that a marked snake on another site had crossed a road.

Corbett (*pers. comm.*) confirms that smooth snakes will cross roads and get killed on them, but they avoid bigger and busier roads and hot tarmac. Sand lizards in Southern England, on the other hand, live in closed populations in limited areas. Individuals which are likely to cross roads will only be those living on adjacent sandy banks. Nevertheless, they seem only to cross small, less busy roads. Common lizards *Lacerta vivipara*, in comparison, disperse more widely, but seem to avoid tarmac. Slow worms are able to maintain permanent populations in a given spot and do not disperse widely, but little is known about their movements. Grass snakes *Natrix natrix*, on the other hand, will travel up to 1.6km or so quite readily. Adders *Vipera berus*, are similarly mobile, but tend to be crushed on wide, busy roads. The same fate has been recorded for a variety of snakes in the Organ Pipe Cactus National Monument in Arizona where high death rates and local depletion in populations were observed rather than any barrier effect (Rosen & Lower, 1994).

Roads are regarded as barriers to amphibians (Reh, 1989 and references therein) but, on the other hand, a great many amphibians cross roads, or attempt to, and are killed. These, apparently conflicting, views are probably related to the scale of the landscape pattern. Animals are killed on local movements to ponds, whilst the barrier effect is on a larger scale and separates and fragments metapopulations. The genetic consequences of this for frogs according to Reh & Seitz (1990) and Reh (1989) are a reduction in the number of heterozygous individuals in a population, increased genetic distancing between populations and a negative influence on the gene flow as they found in a population isolated by roads and a railway for only 30 years (10-12 generations of frogs).

The factors which cause the barrier effect are considered by Mader (1984) to include the following.

- The altered microclimate on the road which would contrast most strongly with that in woodland.

- The emissions, noise disturbance, dust, light and increased salinity of the verges.

- Environmental instability resulting from regular mowing.

- Contrastive habitat on verges.

- The danger to life of crossing roads.

## RESEARCH

Nearly all the research has been centred at the level of the individual, and although these effects are intuitively directly related to effects at the population level (Opdam, 1997), nature conservation significance focuses essentially on this broader scale.  There is very little research which demonstrates significant effects of roads as barriers at the population level.  The clearest are the work on frogs by Reh described above, and the impact found on badgers in the Netherlands (Apeldoorn, 1997).  This lack of evidence may be because of lack of research, or that the multiplicity of variables makes interpretation and analysis particularly difficult, rather than there not being any significant effects for a range of species.  There is also a noticeable dearth of any consideration of plants, and little on birds, although work by Opdam *et al*. (1993) and others can be used to suggest how roads increase the resistance some bird species already face in a fragmented landscape.

## MITIGATION MEASURES

### Avoidance and habitat creation

In particularly sensitive areas, the only adequate measure would be to alter the route of a road to avoid bisecting an important habitat and creating a barrier.  In less valuable sites, it may be possible to extend significantly the affected habitat in other directions and thus provide alternative areas for species.  The main problems with this are the time it takes for a new habitat, particularly involving woody species, to develop, and our limited skill at present in creating new habitats with enough character of the longer established ones to be able to support a wide range of species.

## Pipes, culverts, tunnels and bridges

The mitigation measures more often provided have been a variety of pipes, culverts, tunnels and bridges. These have been very successful in catering for particular species, especially those which can and need to cross, but which get killed crossing roads. There is, however, substantial variation in the form of these connections, and some perform better than others for certain species.

### Pipes

At the small end of the spectrum are pipes connecting each side of the road. A variety from 150 to 400mm in diameter have been used in different schemes in the expectation that some species will find and use them. However, there is no published data on their general effectiveness. However, where combined with funnel shaped entrances, and set just below the surface they can be successful conduits for amphibians (see T. Langton's paper).

Concrete pipes 600mm in diameter (although Douwel (1997) suggests >300mm adequate) have also been successfully used for badgers, but must be combined with secure and constantly maintained badger fencing to prevent access to the road and guide the animals to the tunnels. These need to be set on well-used badger routes to optimise their use. Douwel (1997) considers that badgers can use tunnels like these up to 150m long, and they have even been known to use them as setts.

Measures like these for badgers in the Netherlands are particularly important at the population conservation scale since large numbers of the species are killed on the dense road network (20–25% of the total population) leading to local extinctions (Bekker, 1997). Between 1985 and 1989, 40 tunnels were constructed on roads in the Netherlands as part of the national highway construction programme (Bekker, 1997). Some tunnels were up to 1.2m across. Since 1990, the incorporation of badger tunnels into the national highways programme has become a standard feature, even in areas where badgers are absent but could return. Provisions are also increasingly being made on the non-highway roads by using reported dead badgers as an indicator of local problems (Bekker, 1997). Apeldoorn (1997) describes a GIS based statistical method for estimating the densities of traffic casualties in relation to the surrounding landscape and hence the effectiveness of taking protective measures for badgers

Monitoring of such tunnels shows that badgers use them regularly provided their specification and location are correct. However, little data exists on how effective such tunnels are for other species. Douwel (1997) for example, found beech marten *Martes foina*, polecat *Mustela putorius*, stoat *Mustela erminea*, and fox *Vulpes vulpes*, using badger tunnels, but there are no data on their functioning for other animals (see Table 1).

Similar kinds of tunnels could also be used for otters *Lutra lutra*, but these will need to be placed in quite different locations. It is recommended, for example, that incorporation of dry underpasses adjacent to stream culverts is needed above flood levels, together with otter-proof fencing to prevent access to the road (Anderson, 1994).

### Larger Tunnels and Culverts

Larger tunnels and culverts also provide opportunities for species, largely vertebrates, to pass under roads. For most, exclusion fencing is also needed to keep animals off the road and to guide them into the tunnel. Fencing without tunnels, as proposed on some roads (e.g. parts of the A120 in Essex) to reduce accidents, would only increase the barrier effect (Anderson, 1994).

Larger tunnels designed for agricultural access, streams or footpaths, can also be used by deer and other animals. The less disturbance by humans the more likely they are to be used. Experience in a number of situations has identified measures to optimise use of the tunnel. The floor should be surfaced with sand, soil or other soft material. It should be straight, and as short as possible so that light can be seen at the other end. The surrounding habitat needs to be re-established right up to the tunnel entrance to provide cover, and logs, rubble or stones extending into the tunnel from the outside habitats will encourage greater use by small animals otherwise fearful of meeting predators in open spaces without cover (Anderson, 1994).

There are very few scientific studies on the effectiveness of such tunnels in general, but work by Hunt et al. (1987) supports the guidance provided above. New underpasses were installed under a railway line in New South Wales, Australia, but only feral predators were found to use them.

Passage by native small mammals was not predicted to occur until the native vegetation had re-established around the tunnel entrances. They used existing culverts readily where regeneration had already occurred.

**Table 1.** European vertebrate species found to have passed through or entered pipes, tunnels/culverts and bridges. [sources: se references in text]

| Species | Vernacular Name | Pipes | Tunnels/ Culverts | Bridges |
|---|---|---|---|---|
| *Cervus elaphus* | Red deer | | + | + |
| *Capreolus capreolus* | Roe deer | | + | + |
| *Dama dama* | Fallow deer | | + | + |
| *Sus scrofa* | Wild boar | | + | + |
| *Vulpes vulpes* | Fox | + | + | + |
| *Lutra lutra* | Otter | + | + | |
| *Meles meles* | Badger | + | + | + |
| *Martes foina* | Beech marten | + | + | + |
| *Mustela putorius* | Polecat | + | + | + |
| *Mustela erminea* | Stoat | | + | |
| *Mustela nivalis* | Weasel | + | + | + |
| *Oryctolagus cuniculus* | Rabbit | | + | + |
| *Lepus europaeus* | Brown hare | | + | + |
| *Felis silvestris* | Wild cat | | + | |
| *Genetta genetta* | Genet | | + | |
| *Talpa europea* | Mole | | + | |
| *Apodemus sylvaticus* | Wood mouse | | + | |
| *Clethrionomys glareolus* | Bank vole | | + | |
| *Microtus agrestis* | Short-tailed vole | | + | |
| *Microtus arvalis* | Common vole | | + | |
| *Sorex araneus* | Common shrew | | + | + |
| *Sorex minutus* | Pigmy shrew | | + | |
| *Crocidura russula* | White toothed shrew | | + | |
| *Sorex coronatus* | French | | + | |
| *Sciurus vulgaris* | Red squirrel | | + | + |
| *Erinaceus europaeus* | Hedgehog | | + | + |
| | Lizards | | + | |
| | Amphibians | + | + | |

Furthermore, survey of 17 culverts under roads in Spain with varying levels of traffic and different complexity of vegetation showed no positive avoidance or reluctance to use culverts (Yanes *et al.*, 1995). Most crossings were by small mammals (77%), with woodmice (*Apodemus sylvaticus*) the main species, and shrews *Sorex* spp. at 13%, and rabbits *Oryctolagus cuniculus*, at 10% recorded at lower frequencies. Larger predators, including wildcats *Felis silvestris*, foxes and genets *Genetta genetta*, all used the culverts (Table 1). Only 3 were not used by reptiles, 2 of which were under a motorway. The average use of all vertebrates was 3.8 tracks per culvert per observation day, which for some was not significantly different from track records in distant areas, while for reptiles (mostly lizards) and small mammals it was higher inside the culverts.

Yanes *et al.* (1995) found that use of the culverts was negatively correlated with road width (average 7.5m) and culvert length (average 13.1m) and positively with height, width and openness for small mammals. The use by rabbits and carnivores was negatively correlated with total highway width. The presence of detritus pits to collect debris on streams and thus avoid blocking culverts was a deterrent for many species. However, Yanes *et al.* (1995) conclude that culverts are an effective method of increasing the permeability of highways and therefore of reducing the barrier effect. Yanes *et al.* (1995) found, in a further Spanish study, that the size of culvert is critical. Amphibians preferred wide structures with water in the entrances and inside the culvert, reptiles only used short culverts with a natural substrate and at grade with the adjacent habitat. Small mammals used many different kinds of structure but were inhibited by water at the entrance, although they too prefer a natural substrate. Rabbits were highly selective and avoided passages <150cm in diameter and obstructed views of the far side. In general, mustelids selected structure at grade with the adjacent habitats and within a short distance of shrubby or forest vegetation, while badgers and beech marten prefer those lined with natural materials. Rosell *et al.* (1997) found that foxes only selected underpasses at least 70m wide. Cervids and wild boar *Sus scrofa*, also only used wider (77m wide) structures. Van der Linden (1997) improved on existing wide underpasses by establishing a wall of tree stumps which effectively screened off a cycle way and road and provided a well-structured wildlife passage. Using mark and recapture techniques, and track identification, the improved passage has been monitored. Bank voles, woodmice, short-tailed field vole *Microtus agrestis*, and 15 other species have been detected including mole *Talpa europaea*, weasel *Mustela nivalis*, and polecat. Compared with a pipe (60cm diameter, 50m long), the stump lined passage functioned better as an animal corridor beneath the road.

These studies all relate to terrestrial vertebrates, and stream culverts are only important in the marginal ledges or banks they can also provide for these species. However, streams in culverts can also provide barriers to aquatic organisms, including fish, if not carefully engineered (Bennett, 1991). If culverts are long with a smooth concrete surface, and high flow velocities, some fish species, or young specimens, and a variety of invertebrates may not be able to pass through them except passively downstream. In a study of highway culverts in Montana, Belford & Gould (1989) found that a variety of trout species could swim distances of only 10m with a bottom velocity of 0.96m/s, but 90m at 0.67m/s. It is important to incorporate niches into any culvert to provide resting sites at regular intervals for fish and other species.

An example of this in practice is at Manchester Airport, where a 300m tunnel for the River Bollin made under a new runway is incorporating riffle and pool sequences created using mini-weirs under water and lined with natural riverbed material with logs and boulders anchored along the channel edges to provide refuges (Anderson, 1994a).

### Bridges

In Britain, bridge design for countering barrier effects is in its early days. One of the best examples is on the M25 where it passes through Epping Forest. Here, a substantial cut and cover bridge allows deer and other species to move freely between the two parts of the Forest. However, bridges of various designs have been more widely adopted in the Netherlands, in particular, and other parts of Europe. Berris (1997), Stegehuis (1997) and Scholma (1997) all describe 'ecoducts' designed specifically to carry different species over roads. Keller and Pfister (1997) reviewed the effectiveness of these 'green bridges' for a variety of species. The bridges studied varied from 8 to 200m in width and were typically covered in soil and vegetation, although some incorporated agricultural access tracks. Studies of ground beetles on 4 bridges across a French motorway, which bisected a large forest, showed hardly any use of the bridges by forest carabids and the presence instead of species typical of dry open habitats. However, a grassland covered bridge did provide a suitable habitat and passage for meadow grasshopper *Chorthippus parallelus*, as it connected its preferred habitat areas with an equivalent type. Interestingly, a more widespread grasshopper, *Chorthippus biguttulus*, had a lower probability of reaching the bridge because it utilised a variety of habitats and was not, therefore, channelled across the bridge.

Preliminary results from infrared video surveys show that bridges are used most often by large mammals (Keller & Pfister, 1997). On wide bridges, animals showed less disturbed behaviour than on narrow bridges. Keller & Pfister (1997) recommended a minimum width of about 50m to be most effective, but it is essential that they are well designed and in the most suitable location to cater for the particular species or groups of species for which they are designed. The authors also stress how important it is to establish the same kind of vegetation on the bridges as occurs on either side.

Similar results are provided in other studies. Berris (1997) describes two 50m wide ecoducts added in 1988 after construction, to the A50 Arnhem to Apeldoorn route across the Veluwe heathland. Red deer *Cervus elaphus*, roe deer *Capreolus capreolus*, wild boar, fox and rabbit used the ecoducts straight away, and badgers began crossing from 1994.

Stegehuis (1997) describes an ecoduct built across the A1 motorway in the east of the Netherlands. It has a 60m span, and is 15m wide in its centre, but 30m at the ends to provide funnel shaped entrances. The 1m thick concrete deck is covered by drainage sheets, above which 10cm of loam with pebbles and a root anchoring material are placed which are topped with 50cm loam with at least an 18% organic content. Two metre high wooden acoustic screens were installed at the edges to hide traffic and reduce traffic noise and tied in to the exclusion fencing along the road. The cost was £700,000 in 1992. Roe deer, red squirrels *Sciurus vulgaris*, brown hare *Lepus europaeus*, hedgehogs *Erinaceus europaeus*, rabbits, foxes, marten, stoats, mice and voles all used the ecoduct (Table 1) but the smaller mammals did not use the nearby tunnels.

Compared with underpasses, tunnels, culverts and pipes, green bridges or ecoducts might be expected to cater for more species, including terrestrial invertebrates and plants provided the habitat on them is appropriate. They are also preferred by some species such as deer. However, their dimensions need to be designed with specific species in mind, or to satisfy the most demanding species for which they need to function. In some instances, this results in substantial widths.

## CONCLUSIONS

The assessment presented shows that roads have the potential to increase resistance in the landscape for a variety of species for which they can provide a barrier. The mitigation being adopted so far includes pipes, tunnels, culverts and bridges, some of which are substantial structures. The observations made on the species using these structures suggest that it is mostly the species which, largely, can cross roads but which get killed on them, which benefit from the measures. This can result in increased chances of survival for local populations by removing barriers caused by fenced roads or when populations are low and threatened from road casualties. There would appear to be, however, a significantly greater potential for a wider variety of species including semi-mobile invertebrates and plants, to cross vegetated green bridges than underpasses of different kinds, particularly where they can establish populations in the new habitat and thus link with those on either side.

The conclusion, as Keller & Pfister (1997) suggest, is that these various wildlife passages provide an effective measure to mitigate the barrier effects for some species but that they can never fully compensate for the impact of roads on habitat fragmentation in sensitive areas.

In countries with dense road networks, habitat fragmentation and barrier effects can be substantial. In the Netherlands, there is growing concern about fragmentation, and the Ministry of Transport and Public Works has set several targets to reduce this. By 2000, for example, 40% of the bottlenecks between the national ecological network connecting natural areas and the main road grid are to be removed relative to a 1986 baseline, with a further target of 90% by 2010 (Canters & Cuperus, 1997). Although 'bottlenecks' are to be further defined and suitable measures identified, the kinds of structures described in this paper, as well as new innovative designs, will form part of the palette of measures available to attain these objectives. Moreover, they will be added retrospectively and not just to new schemes. There are lessons here for Britain and other countries.

## REFERENCES

**Anderson P. 1994.** *Roads and nature conservation: Guidance on impacts, mitigation and enhancement.* Peterborough: English Nature.

**Anderson P. 1994a.** *Manchester Airport Second Runway: Proof of Evidence – Nature Conservation.* Manchester Airport plc.

**Apeldoorn van RC. 1997.** Fragmented mammals: What does that mean? In: *Proc. of the International Conference on Habitat Fragmentation, Infrastructure and the role of Ecological Engineering; Maastricht and the Hague, the Netherlands* 17-21 Sept 1995. 121-126.

**Baur A & Baur B. 1990.** Are Road Barriers to Dispersal in the Land Snail Arianta arbustorum? *Canadian Journal of Zoology* **68:** 613-617.

**Belford DA & Gould WR. 1989.** An Evaluation of Trout Passage through six Highway Culverts in Montana. *North American Journal of Fisheries Management* **9:** 437-445.

**Bekker GJ & Canters KJH.1997.** The continuing story of badgers and their tunnels. In: *Proc. of the International Conference on Habitat Fragmentation, Infrastructure and the role of Ecological Engineering; Maastricht and the Hague, the Netherlands* 17-21 Sept 1995.344-353.

**Bennett AF. 1991.** Roads, roadsides and wildlife conservation: a review. In: Saunders D & Hobbs RJ eds. *Nature Conservation 2: The Role of Corridors.* Australia: Surrey Beatty & Sons Pty. Limited. 99-117.

**Berris L. 1997.** The importance of the ecoduct at Terlet for migrating animals. In: *Proc. of the International Conference on Habitat Fragmentation, Infrastructure and the role of Ecological Engineering; Maastricht and the Hague, the Netherlands* 17-21 Sept 1995. 418-420.

**Canters KJ & Cuperus R. 1997.** Assessing fragmentation of bird and mammal habitats due to roads and traffic in transport regions. In: *Proc. of the International Conference on Habitat Fragmentation, Infrastructure and the role of Ecological Engineering; Maastricht and the Hague, the Netherlands* 17-21 Sept 1995. 160-170.

**Corbett KF & Tamarind DL. 1979.** Conservation of the Sand Lizard, Lacerta agilis, by Habitat Management. *British Journal of Herpetology* 5: 799-823.

**Dennis RLH. 1986.** Motorways and Cross-Movements: An insect's 'mental map' of the M56 in Cheshire. *AES Bulletin* 45: 228-241.

**Douwel CK. 1997.** Provisions for badgers realised in the Heumen/A73-area. In: *Proc. of the International Conference on Habitat Fragmentation, Infrastructure and the role of Ecological Engineering; Maastricht and the Hague, the Netherlands* 17-21 Sept 1995.404-408.

**Harris S. 1989.** Taking Stock of the Brock. *BBC Wildlife* 7. London: BBC 460-464.

**Hunt A et al. 1987.** Movement of mammals through tunnels under railway lines. *Australia Zoologist,* 24: 2, 89-93.

**Keller V & Pfister HP. 1997.** Wildlife passages as a means of mitigating effects of habitat fragmentation by roads and railway lines. In: *Proc. of the International Conference on Habitat Fragmentation, Infrastructure and the role of Ecological Engineering; Maastricht and the Hague, the Netherlands* 17-21 Sept 1995. 70-80 & 409-417.

**Linden van der PJH. 1997.** A wall of tree-stumps as a fauna-corridor. In: *Proc. of the International Conference on Habitat Fragmentation, Infrastructure and the role of Ecological Engineering; Maastricht and the Hague, the Netherlands* 17-21 Sept 1995.

**Mader HJ. 1984.** Animal Habitat Isolation by Roads and Agricultural Fields. *Biological Conservation* 29: 81-96.

**Mader HJ et al. 1990.** Linear Barriers to Arthropod Movements in the Landscape. *Biological Conservation* 54: 209-222.

**Merriam G et al., 1989.** Barriers as Boundaries for Metapopulations and Demes of *Peromyscus leucopus* in Farm Landscapes. *Landscape Ecology* 2: 4.227-235.

**Munguira ML & Thomas JA. 1992.** Use of Road Verges by Butterfly and Burnet Populations, and the Effect of Roads on Adult Dispersal and Mortality. *Journal of Applied Ecology* **29**: 316-329.

**Reh E & Seitz A. 1990**. The Influence of Land Use on the Genetic Strueture of Populations of the Common Frog *Rana temporaria*. *Biological Conservation* **54**: 239-249.

**Opdam P *et al.* 1993.** Population Responses to Landscape Fragmentation. In: Vos CC & P. Opdam P eds. *Landscape Ecology of a Stressed Environment*. London: Chapman & Hall. 147-171.

**Opdam PFM. 1997.** How to choose the right solution for the right fragmentation problem? In: *Proc. of the International Conference on Habitat Fragmentation, Infrastructure and the role of Ecological Engineering; Maastricht and the Hague, the Netherlands* 17-21 Sept 1995. 55-60.

**Oxley DJ *et al.* 1974.** The Effects of Roads on Populations of Small Mammals. *Journal of Applied Ecology* **11**: 51-59.

**Reh W. 1989**. Investigations into the Influences of Roads on Genetic Structure of Populations of the Common Frog *Rana temporaria*. In. Langton TES ed. *Amphibians and Roads* ACO Polymer Products Limited. 101-103.

**Reh E & Seitz A. 1990,** The Influence of Land Use on the Genetic Structure of Populations of the Common Frog *Rana temporaria*. *Biological Conservation* **54**: 239-249.

**Rosell C *et al.* 1997.** Mitigation of barrier effect of linear infrastructures on wildlife. In: *Proc. of the International Conference on Habitat Fragmentation, Infrastructure and the role of Ecological Engineering; Maastricht and the Hague, the Netherlands* 17-21 Sept 1995. 367-372.

**Rosen PC & Lowe C.H.** Highway Mortality of Snakes in the Sonoran Desert of Southern Arizona. *Biological Conservation* **68**: 143-148.

**Schonewald-Cox C & Beucher M. 1992.** Park Protection and Public Roads. In: Fielder PL & Jain SK eds. *Conservation Biology: The Theory and Practice of Nature Conservation and Management* London: Chapman & Hall. 374-395.

**Scholma HL. 1997.** A1, Veluwe, Hertaflex. Adaptation of existing viaduct and tunnel to encourage the passage of animals: planned ecoduct. In: *Proc. of the International Conference on Habitat Fragmentation, Infrastructure and the role of Ecological Engineering; Maastricht and the Hague, the Netherlands* 17-21 Sept 1995. 394-396.

**Spellerberg IF & Phelps TE. 1977.** Biology, General Ecology and Behaviour of the Snake, *Coronella austriaca* Laurenti. *Biological Journal of the Linnean Society* **9**: 133-164.

**Stegehuis A. 1997.** Defragmentation Measures in a Small-scale Area of the east Netherlands. In: *Proc. of the International Conference on Habitat Fragmentation, Infrastructure and the role of Ecological Engineering; Maastricht and the Hague, the Netherlands* 17-21 Sept 1995. 387-391.

**Swihart RK & Slade NA. 1984.** Road Crossing in *Sigmodon hispidus* and *Microtus ochrogaster. Journal of Mammalogy* **65:** 2 357-360.

**Yanes M et al. 1995.** Permeability of Roads and Railways to Vertebrates: The Importance of Culverts. *Biological Conservation* **71:** 217-222.

# 17

# Roads and habitat corridors
# for animals and plants

## D G DAWSON

This paper examines theoretical and empirical questions on the extent to which roads constitute a barrier to the movement of animals and plants, whether they assist movement along their length, and on whether conduits across roads can overcome any barrier effects. I examine these questions from a perspective of population ecology.

Much of the background to this paper can be found in a detailed review I prepared for English Nature (Dawson, 1994a; 1995), which examined habitat corridors in a wider context. The conclusion was that corridors should be provided, protected and enhanced, but that they are far from a universal panacea for problems of declining species. Those papers, and other recent work (Bennett, 1991; Hobbs, 1992; Caughley, 1994; Anderson, 1994; Canters, 1997), should be consulted for fuller details of the corridor concept in the context of the problems of declining or small natural populations.

## THEORY ON THE ADVERSE EFFECTS OF BARRIERS

### The faunal collapse theories

The adverse effects of barriers were first theorised by Preston (1962), but it was not till the mid 70s that Diamond (1974, 1975) and others hijacked MacArthur & Wilson's (1963, 1967) theory of island biogeography to develop a theory of faunal collapse (Dawson 1994). This postulated that the fragmentation of habitat leads to the loss of species through two processes: the reduction in the size of habitat patches, and their increased isolation from sources of immigration. The barrier effect of roads is an example of the second of these processes. Kirby (1997) suggests that roads may be greater barriers than their width may suggest, because of their particularly unfriendly habitat.

More recent theorising (Markham 1996, Canters 1997) has concentrated on Levin's (1970) *metapopulation* theory, and has also examined the various synergistic processes that may contribute to the death rattle of the final actor, the *extinction vortex,* including genetic effects (Caughley 1994, Forman *et al.,* 1997).

These classical theories of the causes of extinctions have largely neglected three other considerations (Dawson, 1994a & b).

- We may wish to conserve the occurrence of species locally, quite regardless of whether they may be declining, or threatened with extinction. Habitat corridors may enable a species to be seen locally, or to occur there more often.

- Some individual animals require a large territory to find their requirements for annual survival. Such individuals may be able to find these requirements within a series of connected habitat patches, where they would not without the connections.

- Seasonal migrants must find their way dependably between the habitats that they require each season. Habitat corridors may enable these essential movements.

The last two of these considerations differ markedly from the classical immigration-extinction theories in that their habitat links must enable all candidate individuals to undertake the movement, whereas the odd rare immigrant suffices to prevent extinction in the classical theories. Also, the classical theories apply to all biota, whereas migration and territory are requirements of some animals, but not of plants.

Pimm *et al.* (1988), Bright (1993) and Dawson (1994a) examined factors that may put species at risk of extinction in fragmented habitat. These are as follows.

- Small population.

- Variable population.

- Low reproductive potential.

- Large body size.

- Longevity.

- Poor dispersal ability.

In terms of MacArthur & Wilson's (1967) 'r-K' spectrum, these are "K" species. In terms of Southwood's (1977) 'Habitat templet' they are associated with spatially homogeneous, 'climax' (or 'plagio-climax') natural habitats of long durational stability.

## Theories from landscape ecology

The recent development of landscape ecology is largely a recycling of the theories reviewed above, but does include some innovation. One such is Mader *et al.* (1990) suggestion that linear landscape barriers to movement, such as roads and railways, may reduce the net distances moved by terrestrial animals by leading them off their direct route of travel (see also Opdam, 1997; Forman *et al.,* 1997). Roads may thus increase the 'resistance' of the landscape to movement.

There has been a tendency to combine various fragmentation effects into an all-embracing landscape view, including such matters as barriers, isolation, patch size, total area of habitat and edge effects, but with little attention to habitat quality (Markham. 1996; Bennett, 1997). This has been at the expense of analytical precision.

## Effects of road mortality on numbers

I have found no theoretical elaboration of the effects of road mortality on the population of animals, perhaps because these effects seem obvious, or because the authors concerned equate any death with a loss to the population. The latter is incorrect because of the possibility of compensatory effects (Opdam, 1997). It is important to address the problem of population decline, rather than that of small populations, as it is often too late once the population is at risk of extinction (Caughley, 1994). Help is available in the enormous literature on the regulation of animal numbers (Caughley, 1994; Floyd *et al.,* 1996). It is necessary to evaluate any mortality in the context of population processes, such as age-specific rates of reproduction and mortality, and the areas to which these rates apply. Even where such effects become manifest in the rate of population change, '$r$', it is necessary to identify the 'key factors' affecting population change (Southwood, 1966:1 & 298-308).

The larger the percentage of deaths that is attributable to road mortality, the more likely it is that this mortality may be a key factor with a correspondingly important influence on $r$ (Douwel, 1995). However, even a high percentage of road kills does not necessarily drive the population process; badgers in Britain and moose in Norway are increasing despite substantial mortality on the roads (Kirby, 1997).

Concern with mortality on roads may stem from an animal rights viewpoint, regardless of whether there may be adverse population effects (Kirby, 1997). Such concerns are not the subject of this paper.

## Advocacy of habitat corridors across roads

The provision of conduits for animals to pass under or over major roads is often advocated to avoid mortality (Langton, 1989; Reh & Seitz, 1990; Bennett, 1991b; Harris & Scheck, 1991; Cox, 1993; Anderson, 1994; Bennett, 1997) and less often to ameliorate barriers to movement (Anderson, 1994; Keller & Pfister, 1997; Forman *et al.,* 1997).

## Road verges as habitat corridors

It is widely acknowledged that the vegetation of road verges can provide habitat for animals and plants, and that this can be significant where the surrounds are deficient in other habitat (Way, 1977). As such habitat often exists in strips paralleling the roads, it has also been supposed that these strips may assist the dispersal of animals and plants (Way, 1977; Hansen & Jensen, 1972; Merriam *et al.,* 1989; Bennett, 1991a&b; Hodkinson & Thompson, 1997).

Again, I am not aware of theory developed particularly for this situation, and I suppose that the proponents rely upon general corridor theory (Dawson, 1994).

There has been a number of studies of the transport of plant propagules by or on motor vehicles (Wace, 1977; Hodkinson & Thompson, 1997). While such transport doubtless takes place, it is different in kind from most natural processes of plant and animal dispersal, and is excluded from this paper.

## EMPIRICAL STUDIES OF BARRIERS, DISPERSAL ABILITIES AND POPULATION DECLINE

Modern extinction theories suffer from the same oversimplifications as did the original island model. Harrison (1991) examined supposed cases of decline in metapopulations to conclude that most did not fit the model in that one or more of the following applied:

- The species appeared to be declining for some reason other than being isolated on small habitat fragments, such as degradation of its habitat.

- The habitat composition of the patches varied significantly, so invalidating an assumption of the theory.

- The species was able to persist on a large habitat fragment, so that its immigration to, or extinction on, smaller or poorer patches was not a metapopulation process, leading to population decline.

- The species had such powers of dispersal that a series of habitat patches supported a single large extinction-resistant population, rather than the individual metapopulations of the theory.

Many species appear to fit the last of these categories, notably birds in lowland Europe (Dawson, 1994), with the possible exception of the nuthatch (Opdam, 1991; Verboom *et al.,* 1991; Bellamy *et al.,* 1993). Smith's (1975) robins in apple trees caricature this situation: *"Fluctuations may bring a species to extinction in a system. I can accept a moderate amount of this but not very much. Obviously, extinction is more frequent in smaller systems. At a ridiculous extreme every death or movement creates an extinction at that exact point. Robins become extinct in an apple tree many times each day whenever they fly elsewhere. To me, frequent extinction is a signal that the system under study is not large enough to include the processes being studied. Move the system boundaries out until extinction becomes rare. This ensures that the organising forces of the system lie within the system."*

At the other extreme, some species disperse so poorly that they can practically never cross the gaps between patches of habitat, with or without the aid of habitat corridors (Dawson, 1994). This is so for British plants that are *ancient woodland indicator* species, such as dog's mercury *Mercurialis perennis* (Peterken, 1974; Peterken & Game, 1984) which appear to have poor dispersal powers and which Pollard *et al.* (1974) found to invade hedgerows very slowly. Long-distance dispersal is extremely rare for several butterfly species (Thomas & Harrison, 1992; Thomas *et al.,* 1992) and in some ground beetles (Gruttke, 1997).

For these two categories of species (those where isolation effects appear to be unimportant), the major driving force behind population decline, at least in Europe, is usually habitat loss or degradation, rather than barriers between habitat fragments (see Caughley, 1994 for a review of other possible causes of decline). We should not be diverted from the pressing need to maintain and improve the amount and quality of wildlife habitat for such species by trivial considerations to do with the precise distribution of the habitat in space (Dawson, 1994).

There should, however, be species with intermediate powers of movement that depend upon dispersal to replenish isolated populations that would otherwise go extinct (Dawson, 1994; Anderson, 1994). For these, the size and nature of the barriers is coincident with conditions that affect dispersal markedly. Thomas & Harrison (1992), considered the silver-studded blue butterfly *Plebejus argus* L., in north Wales to be able readily to reach habitat patches within 500m, to encounter increasing difficulty up to a few kilometres and to be unable to reach more distant patches. Thus, at distances between 500 m and a few kilometres, isolation affects the incidence of this species. Similarly Macdonald & Smith (1990) and Bright (1997) considered that British mammals have a range of dispersal abilities such that some are affected by barriers to dispersal. Opdam *et al.* (1993) built this three-way division into their model of population responses to habitat fragmentation.

We need not confine our attention to such species that may be at risk of national or regional extinction, as many people enjoy the presence of susceptible species on a purely local basis.

## ROADS AS BARRIERS

### A barrier to movement

Studies of animals in relation to roads as barriers have established that, for some species, movement may be much more frequent within the habitat on either side than across the road (Oxley, Fenton & Carmody, 1974; Adams & Geis, 1983; Mader, 1984; Merriam *et al.,* 1989; Mader *et al.,* 1990; Lyle & Quinn, 1991; Bennett, 1991a; Munguira & Thomas, 1992; Anderson, 1994; Vos, 1997). Some of these effects are so pronounced as to obviate various methodological and statistical deficiencies in the studies (many of them had inadequate control areas and few applied statistical tests, or calculated confidence intervals). Merriam *et al.* (1989) reviewed some of these studies to reach more detailed conclusions, but these are not supported by the studies' statistical adequacy. The taxonomic groups for which these effects have been claimed are among those for which such effects may be expected (mammals and flightless, or poorly flighted, invertebrates).

Fewer studies report the absence of a barrier effect. However, it is well known that such results are less likely to be published (Csada, 1996), and there are some studies that conclude that the movements of many species are little affected by such features as roads (Duelli, 1990; Dawson, 1994).

Thus the empirical results in relation to roads support those from theory and from a general review of dispersal abilities. So what are the effects of these barriers?

It is nearly impossible for a direct study of movement to establish that rare colonisation events do not occur (the binomial 95% confidence limits for an observation of zero movements are 0 to 3.7). Those who conclude to the contrary (for example Vos, 1997) do not present statistics to defend their position. Thus movement evidence on roads as barriers does not show them to prevent eventual recolonisation nor gene flow: the classical metapopulation concerns. The genetic findings of Reh & Seitz (1990) are often cited to the contrary, however this study provided no evidence of the population effects that would be expected from genetic maladaptation (Dawson, 1994). Kirby (1997) also notes that the postulated genetic effects of isolation by roads have not been demonstrated.

While it is possible that the barriers imposed by a network of roads, in sum, can cause the extinctions of metapopulations of animals and plants (Forman *et al.*, 1997; Vos, 1997), it has proved very difficult in practice to distinguish this barrier effect from other correlated changes, such as the degradation and loss of habitat (Opdam, 1991; Dawson, 1994). For example, Vos & Chardon (1998) were able to account for 65% of the variation in the incidence of moor frogs *Rana arvalis*, in their breeding habitat but, of this, 90% was accounted for by habitat quantity and quality and only 10% by roads. The jury is still out on the supposed general metapopulation or landscape ecology effects of road barriers, but not on the importance of quantity and quality of habitat.

However, a major effect can be expected for those specific animals for which every candidate must be able to cross the barrier and which experience a large barrier effect. It is particularly for these animals that conduits under or over roads have often been provided (Anderson, 1994; Keller & Pfister, 1997). Good links are also required where the local incidence of a species is to be conserved. Some studies do suggest that large numbers of animals can use a good link (e.g. Berris, 1997).

Given this important consideration, it is remarkable that the success of bridges or tunnels to overcome the road barrier appears to be judged universally upon the numbers using the conduit, without any contextual information on the numbers that might be expected to cross the line of the road were the road not there (Forman *et al.*, 1997; Opdam, 1997). Proof that animals use a conduit is decidedly not proof that every candidate to pass the barrier does so successfully, nor is it proof that any adverse effect on the population is mitigated by the conduit.

## Mortality on the road

Together with van Langevelde & Jaarsma (1997), I have not found studies that related mortality on roads to the overall population survival of the species concerned, though statistics like the 45% of all known Florida panther *Felis concolor coryi*, deaths that occurred on highways (Harris & Scheck, 1991) suggest that this may be a significant factor for some species. A review of European studies (Markham, 1996) was unable to place the mortality of most species into a population context, but did suggest that there may be significant concerns for the red fox *Vulpes vulpes* (with an estimated 60% of all mortality on roads), badger *Meles meles* (10-25%), otter *Lutra lutra,* barn owl *Tyto alba,* and amphibia.

Bennett (1991b) reviewed studies of mortality of animals on roads to conclude that, for most species, while the gross numbers dying may be very large, there would be little effect on the population. Even for the species where the mortality may be significant, there were not any studies showing that movement was prevented. Subsequent work (Anderson, 1994; Slater, 1994; Vos, 1997; Mead, 1997) does not contradict these findings. Forman *et al.* (1997) considered that few rare animals in north America are threatened by roads. Some studies (i.e. Dowel, 1997) attribute an increased population to the effects of a conduit provided to pass a road, but without considering alternative explanations for the increase.

Thus it appears that a minority of species may suffer such high mortality on roads that this threatens survival, at least locally. However, there are no good studies the effects of the mortality. Most species probably do not suffer significant population effects.

## ROAD VERGES AS CORRIDORS

Verkaar (1990) studied animals and plants of road verges. Here a decline in species occurrence with distance from the source area has been taken as proof of the utility of the corridor connections. Many of the other studies that have fuelled speculation that road verges promote dispersal have done no more than document occurrences of particular species in the verges (Scott & Davison, 1982; Webster & Juvik, 1983; Scott, 1985) or minor movements there (Vermeulen, 1994). The occurrence of species in suitable habitat, or their decline with distance from a source, is no evidence that corridors got them there. Comparisons should be made with comparable habitat lacking corridor connections, as the species may occur naturally in some places this small or remote, regardless of corridor connections. Other studies failed to establish distance effects (Helliwell, 1975). Clearly the

proposition that road verge habitats may be essential for the dispersal of animals or plants through the landscape requires further study.

The position is different in relation to the requirements of individual animals to migrate or to put together enough habitats for individual survival. Suckling (1984) found that sugar gliders *Petaurus breviceps,* dispersed along a roadside strip of vegetation. Bennett (1990) found that some native Australian mammals move between patches of forest along roadside forest strips. Studies of the habitat requirements of individual animals can suggest a roadside conduit. Saunders & Ingram (1987) found that roadside remnants of natural vegetation acted as corridors for Carnaby's cockatoos *Calyptorhynchus funereus*, breeding in the Western Australian wheatbelt, allowing breeding birds to use remote patches of suitable habitat. Remnant populations died out in the absence of such corridors. Such results suggest that corridors are effective in fulfilling a habitat size threshold for a species constrained by 'central place foraging' (Covitch, 1976; Stephens & Krebs, 1986).

Thus it is possible that road verges allow some animals to commute between the patches of habitat that they need to survive. Clearly these would be those habitat specialists that do not readily cross the matrix between their patchy habitat in the absence of the verge habitat.

## A CAUTION

Scientific works on the effects of roads are typically well qualified by words of warning on their limitations. Nevertheless, there is a danger of oversimplification engendered by road designers seeking cheap, quick and effective mitigation for adverse effects, and scientists under pressure, and fearing a road without mitigating features, replying with more certainty than is prudent. It may be more honest, in our present state of knowledge, to advise that the adverse effects can only be avoided by not having the road. We already have a requirement in Europe for the preservation of habitat corridors and stepping stones (The Council of the European Communities 1992; Article 10), and the development of a 'European Ecological Network' (Bennett, 1997); the scientific basis for these connecting features is not good (Dawson, 1994). Most of the difficulties with conserving nature have nothing to do with habitat connections.

# REFERENCES

**Adams LW, Geis AD. 1983.** Effect of roads on small mammals. *Journal of Applied Ecology.* **20:** 403-15.

**Anderson P. 1994.** *Roads and nature conservation. Guidance on impacts, mitigation and enhancement.* Peterborough: English Nature.

**Bellamy PE, Hinsley SA, Newton I. 1993.** Breeding birds of small woods in an agricultural landscape: an application of island biogeography theory. In: Haines-Young R. ed. *Landscape ecology in Britain.* Department of Geography, University of Nottingham. 35-44.

**Bennett AF. 1990.** Habitat corridors and the conservation of small mammals in a fragmented forest environment. *Landscape Ecology.* **4:** 109-122.

**Bennett AF. 1991a.** What types of organism will use corridors? In: Saunders DA, Hobbs RJ. eds. *Nature Conservation 2: the role of corridors.* Chipping Norton, NSW: Surrey Beatty. 407-408.

**Bennett AF. 1991b.** Roads, roadsides and wildlife conservation: a review. In: Saunders DA, Hobbs RJ. eds. *Nature Conservation 2: the role of corridors.* Surrey Beatty, Chipping Norton, NSW. 99-108.

**Bennett G. 1997.** Habitat fragmentation: the European dimension. In: Canters K. ed. *Habitat fragmentation and infrastructure.* Delft: Ministry of Transport, Public Works and Water Management. 61-69.

**Berris L. 1997.** The importance of the ecoduct at Terlet for migrating mammals. In: Canters K. ed. *Habitat fragmentation and infrastructure.* Delft: Ministry of Transport, Public Works and Water Management. 418-420.

**Canters K. ed. 1997.** *Habitat fragmentation and infrastructure* Delft: Ministry of Transport, Public Works and Water Management.

**Caughley G. 1994.** Directions in conservation biology. *Journal of Animal Ecology* **63:** 215-244.

**Council of the European Communities. 1992.** On the conservation of natural habitats and of wild fauna and flora. *Council Directive 92/43/EEC.*

**Cox PR. 1993.** *Badgers on site. A guide for developers and planners.* Reading: Berkshire County Council.

**Covitch AP. 1976.** Analysing shapes of foraging areas: some ecological and economic theories. *Annual Review of Ecology and Systematics.* **7:** 235-257.

**Csada RD. 1996.** The "file drawer problem" of non-significant results: does it apply to biological research? *Oikos* **76:** 591-3.

**Dawson DG. 1994a.** *Are habitat corridors conduits for animals and plants in a fragmented landscape? A review of the scientific evidence.* English Nature Research Report 94.

**Dawson DG. 1994b.** Road maintenance. *New Scientist.* **143**(1945): 50.

**Dawson DG. 1995.** Narrow is the way. In: Dover J. ed. *Proceedings of the 3rd Annual Conference of IALE(UK).* 30-37.

**Diamond JM. 1974.** Relaxation and differential extinction on land-bridge islands: applications to natural preserves. *Proceedings of 16th International Ornithological Congress.* 616-628.

**Diamond JM. 1975.** The island dilemma: lessons of modern biogeographic studies for the design of natural preserves. *Biological Conservation.* **7:** 129-146.

**Douwel CK. 1997.** Provisions for badgers realised in the Heumen/A73 area. In: Canters K. ed. *Habitat fragmentation and infrastructure.* Delft: Ministry of Transport, Public Works and Water Management. 404-408.

**Duelli P. 1990.** Population movements of arthropods between natural and cultivated areas. *Biological Conservation.* **54:** 193-207.

**Floyd RB, Sheppard AW, DeBarro PJ. eds. 1996.** *Frontiers of population ecology.* Collingwood: CSIRO Publishing.

**Forman RTT, Friedman DS, Fitzhenry D, Martin JD, Chen AS, Alexander LE. 1997.** Ecological effects of roads: towards three summary indices and an overview for North America. In: Canters K. ed. *Habitat fragmentation and infrastructure.* Delft: Ministry of Transport, Public Works and Water Management. 40-54.

**Forman RTT, Godron M. 1981.** Patches and structural components for a landscape ecology. *Bioscience.* **31:** 733-740.

**Gruttke H. 1997.** Impact of landscape changes on the ground beetle fauna (Carabidae) of an agricultural countryside. In: Canters K. ed. *Habitat fragmentation and infrastructure.* Delft: Ministry of Transport, Public Works and Water Management. 145-158.

**Hansen K, Jensen J. 1972.** The vegetation on roadsides in Denmark. A qualitative and quantitative composition. *Dansk Botanisk Arkiv.* **28:**7-61.

**Harris LD & Scheck J 1991.** From implications to applications: the dispersal corridor principle applied to the conservation of biological diversity. In: Saunders DA, Hobbs RJ. eds. *Nature Conservation 2: the Role of Corridors.* Chipping Norton, NSW: Surrey Beatty. 189-220.

**Harrison S. 1991.** Local extinction in a metapopulation context: an empirical evaluation. *Biological Journal of the Linnean Society.* **42:** 73-88.

**Helliwell DR. 1975.** The distribution of woodland plant species in some Shropshire hedgerows. *Biological Conservation.* **7:** 61-72.

**Hobbs RJ. 1992.** The role of corridors in conservation: solution or bandwagon? *Trends in Ecology and Evolution.* **7:** 389-392.

**Hodkinson DJ, Thompson K. 1997.** Plant dispersal: the role of man. *Journal of Applied Ecology.* **34:** 1484-1496.

**Keller V, Pfister HP. 1997.** Wildlife passages as a means of mitigating effects of habitat fragmentation. In: Canters K. ed. *Habitat fragmentation and infrastructure.* Delft: Ministry of Transport, Public Works and Water Management. 70-80.

**Kirby KJ. 1997.** Habitat fragmentation and infrastructure: problems and research. In: Canters K. ed. *Habitat fragmentation and infrastructure.* Delft: Ministry of Transport, Public Works and Water Management. 32-39.

**Langevelde F van, Jaarsma CF. 1997.** Habitat fragmentation, the role of minor rural roads and their traversability. In: Canters K ed. *Habitat fragmentation and infrastructure.* Delft: Ministry of Transport, Public Works and Water Management. 171-182.

**Langton TES. 1989.** *Amphibia and roads.* Shefford, Beds: ACO Polymer Products.

**Levins R. 1970.** Extinction. In: Gerstenhaber M. ed. *Some mathematical problems in biology.* Rhode Island: American Mathematical Society. 75-107.

**Lyle J, Quinn RD. 1991.** Ecological corridors in urban southern California. *Wildlife conservation in metropolitan environments.* Maryland: National Institute for Urban Wildlife.

**MacArthur RA, Wilson E. 1963.** An equilibrium theory of insular zoogeography. *Evolution.* **17:** 373-387.

**MacArthur RA, Wilson E. 1967.** *The theory of island biogeography.* Princeton, New Jersey: Princeton University Press.

**Markham D. 1996.** *The significance of secondary effects from roads and road transport on nature conservation.* English Nature Research Report 178.

**Merriam G, Kozakiewicz M, Tsuchiya E, Hawley K. 1989.** Barriers as boundaries for metapopulations and demes of *Peromyscus leucopus* in farm landscapes. *Landscape Ecology.* **2:** 222-235.

**Macdonald DW, Smith H. 1990.** Dispersal, dispersion and conservation in the agricultural ecosystem. In: Bunce RGH, Howard DC. eds. *Species Dispersal in agricultural habitats.* London: Bellhaven. 18-64.

**Mader H-J. 1984.** Animal habitat isolation by roads and agricultural fields. *Biological Conservation.* **29:** 81-96.

**Mader H-J, Schell C, Kornacker P. 1990.** Linear barriers to arthropod movements in the landscape. *Biological Conservation.* **54:** 209-222.

**Mead C. 1997.** Pathetic bundles of feathers - birds and roads. *British Wildlife.* **8:** 229-232.

**Munguira ML, Thomas JA. 1992.** Use of road verges by butterfly and burnet populations, and the effect of roads on adult dispersal and mortality. *Journal of Applied Ecology.* **29:** 316-329.

**Opdam P. 1991.** Metapopulation theory and habitat fragmentation: a review of Holarctic breeding bird studies. *Landscape Ecology.* **5:** 93-106.

**Opdam PRM. 1997.** How to choose the right solution for the right fragmentation problem? In: Canters K. ed. *Habitat fragmentation and infrastructure.* Delft: Ministry of Transport, Public Works and Water Management. 55-60.

**Opdam P, Apeldoorn R van, Schotman A, Kalkhoven J. 1993.** Population responses to landscape fragmentation. In: Vos CC, Opdam P. *Landscape ecology of a stressed environment*. London: Chapman & Hall. 147-171.

**Oxley DJ, Fenton MB, Carmody GR. 1974.** The effects of roads on populations of small mammals. *Journal of Applied Ecology.* **11:** 51-59.

**Peterken GF. 1974.** A method for assessing woodland flora for conservation using indicator species. *Biological Conservation.* **6:** 239-245.

**Peterken GF, Game M. 1984.** Historical factors affecting the number and distribution of vascular plant species in the woodlands of central Lincolnshire. *Journal of Ecology.* **72:** 155-182.

**Pimm SL, Jones HL, Diamond J. 1988.** On the risk of extinction. *The American Naturalist.* **132:** 757-785.

**Pollard E, Hooper MD, Moore NW. 1974.** *Hedges.* London: Collins.

**Preston FW. 1962.** The canonical distribution of commonness and rarity. *Ecology.* **43:** 185-215.

**Reh W, Seitz A. 1990.** The influence of land use on the genetic structure of the common frog *Rana temporaria*. *Biological Conservation.* **54:** 239-249.

**Saunders DA, Ingram JA. 1987.** Factors affecting survival of breeding populations of Carnaby's cockatoo *Calyptorhynchus funereus latirostris* in remnants of native vegetation. In: Saunders DA, Arnold GW, Burbridge AA, Hopkins AJM. eds. *Nature conservation: the role of remnants of native vegetation*. Chipping Norton, NSW: Surrey Beatty. 249-258.

**Scott NE. 1985.** The updated distribution of maritime species on British roadsides. *Watsonia.* **15:** 381-386.

**Scott NE, Davison AW. 1982.** De-icing salt and the invasion of road verges by maritime plants. *Watsonia.* **14:** 41-42.

**Skinner C, Skinner P, Harris S. 1991.** An analysis of some of the factors affecting the current distribution of badger *Meles meles* setts in Essex. *Mammal Review.* **21:** 51-65.

**Slater F. 1994.** Wildlife road casualties. *British Wildlife.* **5:** 214-211.

**Smith FE. 1975.** Ecosystems and evolution. *Bulletin of the Ecological Society of America.* **56:** 2.

**Southwood TRE. 1966.** *Ecological methods.* London: Methuen.

**Southwood TRE. 1977.** Habitat, the templet for ecological strategies? *Journal of Animal Ecology.* **46:** 337-365.

**Stephens DW, Krebs JR. 1986.** *Foraging Theory.* Princeton, N.J: Princeton University Press.

**Suckling GC. 1984.** Population ecology of the sugar glider, *Petaurus breviceps*, in a system of fragmented habitats. *Australian Wildlife Research.* **11:** 49-75.

**Thomas CD, Harrison S. 1992.** Spatial dynamics of a patchily distributed butterfly species. *Journal of Applied Ecology.* **61:** 437-446.

**Thomas CD, Thomas JA, Warren MS. 1992.** Distribution of occupied and vacant butterfly habitats in fragmented landscapes. *Oecologia.* **92:** 563-567.

**Verboom J, Schotman A, Opdam P, Metz JAJ. 1991.** European nuthatch metapopulations in a fragmented agricultural landscape. *Oikos.* **61:** 149-156.

**Verkaar HJ. 1990.** Corridors as a tool for plant species conservation? In: Bunce RGH, Howard DC. eds. *Species dispersal in agricultural habitats.* London: Belhaven. 82-97.

**Vermeulen HJW. 1994.** Corridor function of a road verge for dispersal of stenotopic heathland ground beetles, Carabidae. *Biological Conservation.* **69:** 339-349.

**Vos CC. 1995.** Effects of road density; a case study of the moor frog. In: Canters K. ed. *Habitat fragmentation and infrastructure.* Delft: Ministry of Transport, Public Works and Water Management. 93-97.

**Vos CC, Chardon JP. 1998**. Effects of habitat fragmentation and road density on the distribution pattern of the moor frog *Rana arvalis. Journal of Applied Ecology* **35:** 44-56.

**Wace N. 1977.** Assessment of dispersal of plant species - the car-borne flora in Canberra. *Proceedings of the Ecological Society of Australia.* **10:** 167-186.

**Way JM. 1977.** Roadside verges and conservation in Britain: a review. *Biological Conservation.* **12:** 65-74.

**Webster L, Juvik JO. 1983.** Roadside plant communities on Mauna Loa, Hawaii. *Journal of Biogeography.* **10:** 307-316.

# 18

# A decision support system for roadside deadwood

## L WYATT

Roadsides in England contain significant areas of original woodland and tree planting which may already, or in the future, provide a positive deadwood resource, thereby contributing to biodiversity. Problems often occur in consistently assessing the deadwood potential and best nature conservation management of an area, in balance with the requirements of road user safety, those living adjacent to the road, and accessibility for maintenance purposes.

## THE SYSTEM

A Decision Support System was developed for possible use within the Highways Agency, the Government body responsible for trunk roads and motorways in England.

### Research

The System was developed from a review of relevant literature, personal experience and the results of questionnaires to those managing deadwood both on roads and elsewhere.

It was then field tested against pre-determined criteria to see whether it could be used by ecologists or site managers to assess the suitability of a roadside site for deadwood management and suggest possible deadwood management techniques.

A literature review revealed that there has been no specific consideration of the importance of roadside deadwood as an ecological resource. Research in the UK has concentrated on ancient or veteran trees and other habitats for species dependent on deadwood. The review also highlighted the complexity of factors that affect the success of deadwood as a habitat; and the importance of deadwood as a ecological resource to a wide range of species.

A questionnaire examined current deadwood management within the Highways Agency and within nature conservation, forestry and other transport organisations with an estate management role.

From this information the System was designed so that it could be easily applied by site managers and workers, who may not have detailed knowledge about deadwood ecology.

## Field trials

The System consists of an assessment based on thirteen characteristics of a site; a comparison of the assessment against appropriate management options; a prompt for recording and reviewing decisions; and guidance notes.

The System was field-tested using three officials from the Highways Agency, on twenty-six sites along the M4 Motorway and A36 Trunk Road in southern England in June 1997. The sites included areas where deadwood management would be both appropriate and inappropriate, and included sites to isolate assessor variance.

## Results

The results were tested against four criteria, namely the System's ability to assess the range of sites tested, to provide a consistent approach, to produce realistic management practices options and to be easy to use.

The System was successful in meeting three of the four criteria; but not so successful in providing a consistent approach. The consistency was affected by assessor variance; problems of assignment of categories; the way the criteria assumed the testers had the same knowledge of deadwood as the System; and the assumption in the System that existing management was carried out using the same logic as the System.

## CONCLUSION

The research has shown that the System could assist site staff and ecologists in assessing roadside sites. Where different constraints affect the roadside the System could form the basis of a similar system with the characteristics changed to reflect any differences.

Due to the limitations of time and resources this research did not include an assessment of the value of the roadside deadwood, therefore monitoring of whether or not the value of deadwood has improved should be carried out on sites where the System has been applied.

The System, subject to further testing, could be used for other applications such as those listed below.

- Designing roadside planting, using a number of the characteristics to assist finding suitable sites and maximising their deadwood potential.

- Training site staff and ecologists on what influences the value of deadwood.

- Developing systems for specific species if sufficient is known of their autecology.

## ACKNOWLEDGEMENT

The support of the Highways Agency, and the assistance of those who provided information and advice is gratefully acknowledged

[This research was undertaken as part of a MSc degree in 'Ecology and Management of the Natural Environment' at University of Bristol, UK, 1997.]

# 19

# Are roads harmful or potentially beneficial to butterflies and other insects?

## J A THOMAS, R G SNAZELL & L K WARD

The aim of this paper is first to examine whether evidence exists to support the popular perception that roads and traffic are detrimental to local populations of insects, concentrating mainly on butterflies, which have been the focus of various campaigns to oppose new road schemes. In the final section we suggest how some potentially harmful impacts of road construction may be mitigated and how, in some circumstances, their design may enhance rather than reduce the diversity of insects in a region. A résumé is given of the two most successful schemes of mitigation concerning insects in the UK – resulting from the construction of the M40 near Oxford and of the M3 through Twyford Down near Winchester which, contrary to popular perceptions, have resulted in substantial net gains in wildlife in each locality.

## BACKGROUND

Despite studies showing that busy roads form a partial barrier to dispersal for certain Carabid beetles (Mader, 1984) and that the bodies of some roadside invertebrates contain high concentrations of lead and zinc (Wade *et al.*, 1980), it has long been apparent from the scientific literature that the central reservations and verges of motorways and major roads in Europe and North America often support high densities of both common and local species of insect from a wide range of taxa, including Coleoptera (beetles), syrphids (hoverflies), aculeate Hymenoptera (ants, bees, wasps), aphids, Auchenorrhyncha (bugs) and moths (Free *et al.*, 1975; Feltwell & Phillip, 1980; Port & Thompson, 1980; Port & Spencer, 1987). Indeed, certain aphids and moths experience periodic population explosions to become pests on the shrubs adjoining roads, a circumstance attributed to enhanced nitrogen levels in foodplants growing near dense traffic (Port &

Thompson, 1980). Nevertheless, many conservationists believe that roads and traffic are inimical to butterflies (e.g. Anon, 1988), with the result that from the late 1970s to the early 1990s in the UK, butterflies were perhaps employed more than any other group of wildlife by protesters opposing new constructions. To our knowledge, the routes of three proposed motorway extensions were altered during that period, mainly because of the presumed harm they would cause to butterflies, particularly the black hairstreak *Strymonidia pruni* (M40, M1) and the chalkhill blue *Lysandra coridon* (M3) (Thomas & Munguira, 1991 and unpublished; Snazell, 1998).

Butterflies were used as 'flagships' to oppose road schemes probably for three reasons: they have an image of vulnerability and fragility; they are the most conspicuous and popular group of insects; yet – in contrast to some other insects and much other wildlife – few data existed prior to 1990 on the effect of roads on their populations. Moreover, an early study involving the orange tip *Anthocaris cardamines* suggested that the M56 was a barrier to the dispersal of this normally mobile butterfly (Dennis, 1986). The extension of this conclusion from a study of a single species at one site to butterflies and roads generally has been criticised on the grounds that the construction of the M56 involved blocking a valley, and that it was not the road or increased traffic *per se.* that obstructed the butterfly (Munguira & Thomas, 1992). In contrast, our initial observations indicated that although most roadsides were poor places to observe butterflies, certain verges appeared to support large populations of common, local and (occasionally) rare species, and that road verges and central reservations were among the few places where any non-migratory butterfly could be found in thè most intensively farmed landscapes of Europe. Moreover, we had often observed busy roads being crossed by species known to have comparatively sedentary adults which, unlike the orange tip, live in discrete 'closed' populations (*sensu* Thomas, 1984). Therefore in 1989, we made a more scientific assessment of the impact of roads on butterflies (Thomas & Munguira, 1991; Munguira & Thomas 1992); the main results are summarised in section 2.

Before describing the conclusions it is worth listing the three main hypotheses about why roads are considered inimical to butterflies.

- New roads destroy butterfly habitats.

- Roads form a barrier to dispersal causing genetic deterioration through inbreeding or an unstable metapopulation structure, both resulting in the extinction of isolated populations.

- Traffic increases mortality through car strike and toxins.

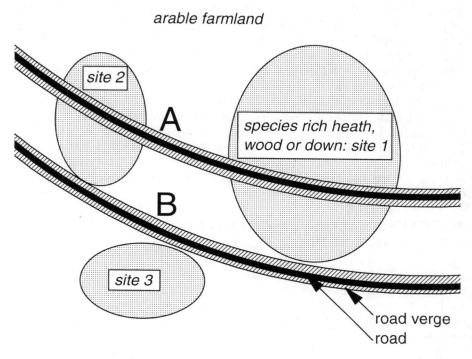

**Figure 1.** Two hypothetical routes (A and B) for a new road through a landscape of intensively farmed land containing three 'islands' of semi-natural biotope.

It is also worth introducing, for later reference, two hypothetical cases of a new road construction, since there has been much confusion about the situations in which each of the above hypotheses apply. Each example runs through a typical lowland landscape (Figure 1) of intensively farmed land (arable and 'improved' grassland) containing three 'islands' (sites 1-3) of SSSI-quality semi-natural biotope, for example heathland, woodland or an unfertilised chalk downland. The farmland typically supports a very low diversity of butterflies (and other wildlife) whereas sites 1-3 are rich in species (Thomas, 1983, 1984). Two possible routes are shown. Route A cuts through sites 1 and 2, destroying a proportion of their valued habitats and, for some species, perhaps dividing the surviving areas into two isolated sites north and south of the road; an alternative hypothesis is that the new road verges (shown hatched in Figure 1), if suitably designed, may provide a better conduit than the former arable land for dispersal between sites 1 and 2. Route B causes no direct loss of sites 1-3, although traffic noise and fumes might pollute them. On the contrary, given suitable verges, route B might provide additional breeding areas for a range of species that are unable to inhabit arable land as well as enhancing the dispersal between sites 1, 2 and 3 of species that are reluctant to cross intensive farmland. In sections 3 and 4, we consider whether the net effects of road schemes such as those depicted in Figure 1 are invariably harmful, neutral, or potentially beneficial for butterflies.

## DO ROADS INVARIABLY HARM BUTTERFLIES AND OTHER INSECTS?

The detailed methodology and results of our study of roads and butterflies are described by Munguira and Thomas (1992). In brief, twelve roads in Dorset and Hampshire were selected, ranging from minor routes to major dual carriageways and trunk roads carrying similar densities of traffic to motorways.

Study sites were selected with no prior knowledge of the number of species associated with them, and encompassed examples combining a range of variation in seven different factors thought likely to affect butterflies. The density of every species of butterfly and burnet moth (Zygaenidae) observed on the verges of each roadside was measured by making weekly standardised transect counts (Pollard *et al.,* 1973) from early June to mid September (14 weeks), and the following variables were measured on each site:

- Topography, classified into three categories of flat/uniform; verges with a ditch; and verges with a bank, slope or uneven terrain.

- Whether the land adjoining the road and its verge consisted of semi-natural biotopes rich in species (e.g. woods, heath or unfertilised grassland), or of intensive agricultural or urban ground where few insects breed.

- Whether the roadside was sheltered by a hedgerow or not.

- The density of traffic measured at midday.

- The number of nectar sources for adult butterflies and burnets on the roadside.

- The width of the verge.

- The number of different breeding habitats for immature butterflies and burnets that were present on road verges.

The last factor is the most important determinant of the diversity and abundance of butterflies in semi-natural ecosystems, and reflects the fact that although the requirements of most adult butterflies are comparatively catholic, those of their young stages (particularly the larvae) are considerably more specialised than had previously been thought in nearly every European species that has been studied. Not only are the larval foodplant(s) often more restricted than those listed in popular books but they must often be growing in abundance in a particular growth-form, microhabitat or microclimate if they are to be used by a particular species (Thomas, 1991).

Three measures of butterfly and burnet abundance – the number of species, number of individuals, and the Shannon-Weaver diversity index – were compared with the seven road characteristics. Additional studies of the size of butterfly populations, the aptitude of adults to cross roads, and of mortality through car strike were made using mark-recapture techniques on the verges of the two busiest roads.

## Can butterfly populations coexist with roads?

Our results supported conclusions from studies of other insects by showing unequivocally that certain road verges contain an abundance and diversity of butterflies, although most do not. One site had 23 species of butterfly and two burnets recorded along the 100 metre transect, and the 12 roadsides contained a total of 27 different species, representing 56% of all butterfly species inhabiting Hampshire and Dorset. This was an un-expectedly high total in view of the fact that recording was restricted to the summer months and none of the roadsides contained or was bordered by woodland. Most butterflies recorded were common species, but some scarce or local species were recorded, including Essex skipper *Thymelicus lineola*, brown argus *Aricia agestis*, grayling *Hipparchia semele*, marbled white *Melanargia galathea*, silver-studded blue *Plebejus argus* and small blue *Cupido minimus*. All but the silver-studded blue had populations that were confirmed or appeared to breed on these roadsides, and in several cases populations were entirely supported by the road verge.

One would not expect to encounter many rarities on twelve randomly-chosen 100 metre transects in any biotope, but more selective searching by ourselves and others (e.g.. Morton, 1985; Ravenscroft, 1994) has revealed populations of additional rare or local species on roadsides, often existing as self-contained colonies. These include the black hairstreak *Strymonidia pruni*, brown hairstreak *Thecla betulae*, Adonis blue *Lysandra bellargus*, chalkhill blue *Lysandra coridon*, grizzled skipper *Pyrgus malvae*, dingy skipper *Erynnis tages*, chequered skipper *Carterocephalus palaemon*, and Lulworth skipper *Thymelicus acteon*, in the UK, and the large blue *Maculinea arion*, Rebeli's large blue *Maculinea rebeli*, Glanville fritillary *Melitaea cinxia*, and heath fritillary *Mellicta athalia* breeding on the verges of major roads in France and Spain.

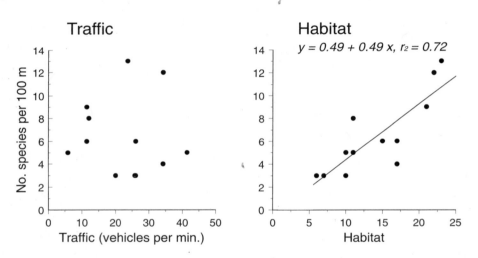

**Fig. 2** The diversity of butterflies and burnets recorded on roads verges compared with traffic volumes and the diversity of larval habitats present

Several populations of common and local butterflies on road verges in Dorset and Hampshire were also exceptionally large (Munguira & Thomas, 1992). On the other hand, most roadsides contained modest numbers (average nine) of species of butterfly, and some were extremely poor. We therefore examined why some road verges contained a butterfly fauna comparable to that of a designated conservation area whereas most others ranged from average for a semi-natural biotope to poor.

## The quality of the neighbouring land

Surprisingly, no measure of butterfly richness (number of species, number of individuals, Biodiversity index) was correlated with the quality of the land adjoining the road and its verge. Indeed, butterfly richness was greatest on roadsides running through urban and intensively farmed land, which were almost bereft of butterflies. In other words, the butterflies recorded beside different roads consisted of colonies that were largely or wholly supported by the road verge itself, rather than vagrants from richer sites nearby. This conclusion was supported by counts of eggs and caterpillars of the more conspicuous species found on verges.

## The density of traffic

The amount of traffic had no apparent affect on the abundance of butterflies on road verges. Indeed, slightly (but not significantly) higher numbers were found beside the busiest roads (Figure 2a), and the density of individuals and the number of species recorded on the central reservation of the busiest road sampled were exceeded, respectively, by only two and four of the other 11 verges.

## The nature of the road verge

The variety of habitats present suitable for the young stages of butterflies and burnet moths was much the most important factor explaining variation in both species-richness (72% of variation; Figure 2b) and the diversity (51%) of butterflies on the twelve roadsides that we examined. In addition, wider verges contained significantly higher *densities* of adult butterflies per square metre, over and above the greater numbers they supported simply because of their greater width. Road verges with an abundance of the flowers for adult feeding also held a greater diversity of butterflies and burnets, although this factor was not statistically significant when account was taken of the variety of breeding habitats that were present (i.e. a flower-rich verge with few breeding places contained fewer butterflies that one containing few flowers but many breeding places). Several other factors were significantly correlated with variation in the number of different breeding habitats present, including the width and topography of sites. Thus wider verges with scattered shrubs and more rugged terrain (especially cuttings) and ditches supported the most butterflies, but only because the sites with these attributes contained the widest variety of habitats for the caterpillar stage.

## Are roads and verges a barrier or conduit for dispersal?

A total of 397 adults belonging to 21 species of butterfly or burnet (listed in Munguira & Thomas, 1992) were seen or detected as flying across the two widest and busiest roads studied in Hampshire and Dorset. Crossings were not detected in another nine species, but these occurred at low densities on the verges. The recapture points of individuals of two species (marbled white *Melanargia galathea* and meadow brown *Maniola jurtina*) are shown for two sites in Figure 3. In most species, neither sex was more likely than the other to cross roads. More recent marking experiments indicate that 2-7% of chalkhill blue populations cross the M3 at Twyford Down.

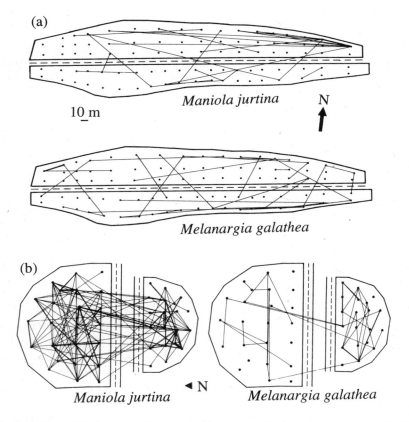

**Figure 3.** Movements made by marked adults of the meadow brown and marbled white butterflies along the verges and across busy trunk roads at (a) Bere Regis (A31) and (b) Poole (A35) in Dorset (from Munguira & Thomas, 1992)

These results indicate that there is little or no chance that butterfly or burnet populations separated by a road will become sufficiently isolated as to experience inbreeding depression. However, in some species we studied and on some – but not all – roads, there was significantly less movement across roads compared to the flight of adults along comparable distances of verges. For example, the meadow brown was less likely to cross the road on both sites illustrated than to fly the same distance along their verge, but there was no significant difference for the marbled white at Bere Regis (Figure 3). At present, no model exists to predict whether these recorded reductions in movement will have any impact on the long-term persistence of metapopulations in a landscape. However, it is possible to deduce whether a new road construction might be harmful or beneficial to metapopulations by reference to the two scenarios in Figure 1.

**Table 1.**

| Species | Common name | Continuous | Roads | Arable |
|---|---|---|---|---|
| *Maniola jurtina* | Meadow brown | 46-47% | 17-21% | 16% |
| *Melanargia galathea* | Marbled white | 45-50% | 10-32% | ? |
| *Lysandra bellargus/* *Lysandra coridon* | Adonis/ chalkhill blue | 2-7% | 1-2% | 0% |
| *Pyronia tithonus* | Gatekeeper | ? | ? | 5% |

There has been much confusion about whether roads represent barriers to dispersal, primarily because account is seldom taken of the character of the land that was present before the road was built. Very few data exist for the dispersal of butterflies across alien (e.g. arable) land for a species, in contrast to its dispersal over continuous areas of optimum habitat. Table 1, compares movements over similar distances, including across roads, for five butterflies. This suggests that roughly half the number of adults is likely to cross a (busy) road than fly the same distance across a continuous patch of suitable habitat, but that 30-50 m. of arable land provides a slightly greater barrier to dispersal than a motorway or major road. Marking experiments of other species that live in closed populations show that very low numbers leave their breeding sites to cross farmland (e.g. Thomas, 1991; Dover, 1991). The effect of a new road on dispersal is therefore likely to depend (i) on the nature of the land where it is built, and (ii) on whether the new verge contains the breeding habitat of a particular species (see above), or simply sheltered flower-rich grassland or hedgerows between breeding sites along which adults of some species are more likely to fly (e.g. Warren, 1987; Sutcliffe *et al.*, 1997).

In our hypothetical landscape (Figure 1), the populations of most local species of butterfly requiring semi-natural habitats are isolated from each other by several hundred metres of arable land. The construction of route B is likely to be wholly beneficial to dispersal, because the short distance across the road is no greater (probably less) of a barrier than the arable ground it replaced, whereas the verges linking sites 1, 2 and 3 may be three or more times as readily traversed than the original farmland, provided they contain suitable habitat for the species in question. Route A has a similar potential to enhance – and is equally unlikely to hinder – flights between sites 1 and 2. However, it may roughly halve the previous movements across the 30-50 m strip lost to the construction, dividing the northern and southern halves of sites 1 and 2.

## Does traffic kill significant numbers of butterflies?

No measurement has been made of indirect effects of traffic (or road materials) on butterflies. However, the absence of a negative correlation between traffic density and either butterfly numbers or diversity (Figure 2a) on both new and long-established verges suggests that any possible reduction in fitness directly from fumes or from an accumulation of chemicals in the soil is insufficient to depress populations. In addition, chalkhill blue *Lysandra coridon* populations increased at Twyford Down in the first three years following the opening of the M3. These results are consistent with studies of other insect taxa (see 'Introduction'). However, there is no evidence that increases at Twyford or the exceptionally high densities of other butterflies on road verges in Hampshire and Dorset (Munguira & Thomas, 1992) reflect an improvement in their habitats caused by enhanced nitrogen from fumes entering larval foodplants (Port & Thompson, 1980).

By the same argument, there is no *a priori* case for suggesting that deaths by car strike significantly reduce butterfly populations. Direct estimates of the proportions of populations killed by vehicles are given by Thomas & Munguira (1992), and range from 0.6%-1.9% for species that live in closed colonies to 7% of the roadside 'population' of the migratory small white *Pieris rapae*. These mortalities are insignificant for insects whose population densities are partly determined by density-dependent factors (Thomas *et al.*, 1994). We suspect that these surprisingly low figures reflect the aerodynamics of modern vehicles. We frequently watched butterflies being swept up and over speeding cars, then fly on safely once the turbulence has passed: in contrast, the radiator grills of the vehicles in the 1950s made fine collections of butterflies and other insects.

## HABITAT CREATION BESIDE ROADS

### General principles

It follows from the previous sections that new roads, including motorways, need not necessarily harm butterflies and other insects; on the contrary, if carefully designed, their verges might enhance the populations of rare, local and common species in a region both by providing additional breeding sites and (probably) by increasing movements between isolated breeding areas. However, most roads probably have neutral or mildly beneficial effects on insect populations: few fulfil their potential as habitats and conduits.

For road schemes fully to enhance insect populations it is essential for insect habitats to be included in the very earliest designs (Bickmore, 1992), when their incorporation is often simple to achieve and involves little or no extra cost. In practice this seldom occurs, as engineers and landscape designers – though increasingly skilled at creating suitable conditions for vertebrates and for a rich and characteristic flora (Wells & Bayfield, 1990) – have less knowledge of the additional requirements of invertebrates; furthermore, ecologists who might advise are often occupied with opposing schemes at this early stage of proposals.

It is beyond the scope of this paper to give comprehensive guidelines to encourage insects, but a few principles are outlined. As with all wildlife, the choice of route is usually important. In general, routes through species-rich semi-natural habitats are likely to result in a net loss of insect diversity whereas those through intensive farmland, urban areas and other poor biotopes are likely to result in a net gain. Routes A and B (Figure 1) represent two extremes in a continuum. For insects, route B is ideal because not only is it confined to arable land but it passes so close to three previously isolated semi-natural biotopes that it has the potential to enhance species' dispersal.

Of course there may be other objections to route B on wildlife grounds: it may result in significant deaths through car strike of certain vertebrates, and may reduce naturalists' enjoyment of sites 1-3. However, these considerations – and those of cost, feasibility, and the impact of the construction on the physical landscape and on archaeological sites – are beyond the scope of this paper. In practice, many new routes fall between the extremes represented by routes A and B. In these cases measures may be possible to mitigate the damage.

Having chosen the route, the design of roadside verges is paramount to their success both as habitat and conduit for insects. The most important single factor is to ensure that the verge contains the greatest possible variety of the specific habitats of the immature stages of the characteristic insect species for those soils in the region. UK butterfly habitats are defined by Thomas (1986, 1991) and Thomas & Lewington (1991), and their incorporation into road designs by Morris et al., (1994). Briefly, in addition to including an abundance of larval foodplants of local provenance, verges should be as wide as possible, with broken or undulating terrain covered usually by a range of thin to skeletal depths of unfertilised local soil with deeper-soiled patches adjoining them; they should also contain a mixture of native scrub (or hedgerows), and have maximum shelter.

Many sites require management by mowing in spring or autumn (Munguira & Thomas, 1992), although annual mowing may not be desirable on thin-soiled verges where plant succession is slow. Deep cuttings, especially those with southerly-facing aspects made through calcareous formations in southern Britain, provide the maximum potential for butterflies and should be given priority in habitat creation schemes. However, a wide variety of microhabitats and shelter can be created on flat ground, by constructing an undulating terrain of sinuous parallel banks and ditches, as illustrated in Figure 4 (from Morris et al., 1994), or simply by leaving the surface broken with many small pits and hummocks. Most new roadsides have exactly this latter structure during construction, but are later flattened into smooth surfaces before being coated with a thick deposit of rich topsoil and sown with cultivars or agricultural ley grasses; all three activities reduce a verge's capacity to support insects.

Two recent schemes in which several of these principles were applied are briefly described below.

## Habitat creation beside the M40

Although most land on the heavy clays north-east of Oxford is intensively farmed, plans to route the M40 between Waterstock and Wendlebury posed a potential threat to various flower-rich meadows associated with Otmoor and to about 700 ha of ancient woodlands around Bernwood Forest, encompassing eight SSSIs. The 1985 published route resembled route B (Figure 1) and 'was just able to thread its way around all the designated sites, even though this resulted in a less than direct route' (Bickmore, 1992). However, the line was then moved slightly following

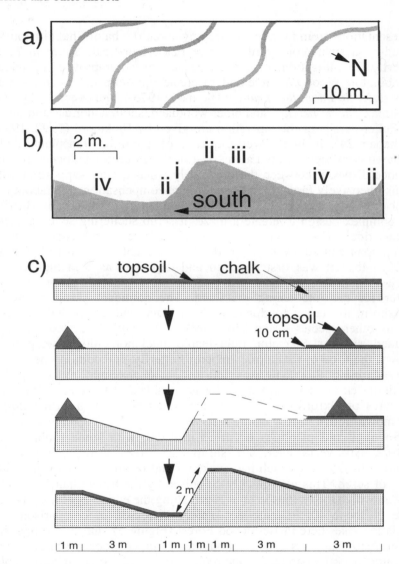

**Figure 4.** Diagrams of 'ridge-and-furrow' terrain to create the maximum diversity of habitats for calcareous grassland butterflies in the UK on formerly flat ground. (a) Aerial map of ridges and furrow. (b) Crossection of one ridge and furrow (key to scarce species' habitats on chalk: i) arid warm turf with much bare chalk (*Hesperia comma, Hipparchia semele, Cupido minimus*); ii) sparse warm dry turf (*Lysandra bellargus, L. coridon, Aricia agestis, C. minimus*); iii) short well-drained turf (*L. coridon, A, agestis, C. minimus*); iv) cooler deeper turf (*Hamearis lucina, Argynnis aglaja*). (c) Crossection showing four stages in the construction and contouring of one ridge and furrow (from Morris *et al.*, 1994).

representations from farmers, destroying about 0.5 ha of Shabbington Wood including a small colony of the rare black hairstreak *Strymonidia pruni* butterfly. In mitigation, over 4 ha of arable and improved ley adjoining Shabbington Wood was released to create optimum habitat for the black hairstreak *Strymonidia pruni* (Thomas, 1975), the very local brown hairstreak *Thecla betulae*, and other woodland, scrub and grassland insects.

Fuller details of this operation are given by Bickmore (1992, 1997 and in chapter 24). In brief, over a quarter of a million transplants and 600 feathered trees were planted along or near the 19 km route, one aim being to enhance dispersal between the previously isolated sites supporting wildlife in this intensively farmed landscape. In the compensation area, about a third of the total area was planted with more than ten thousand shrubs, distributed in a complex design to create a maze of scrub sheltering seven glades and various rides. This provided 3.3 km of sheltered shrub-edge, with many indentations and sunny bays. In the former arable portion (c 2.5 ha), the enriched topsoil was first removed and deposited as 1 m high banks on which the shrubs (of local provenance) were planted, providing further shelter for the intervening grassland. The dominant shrub planted was blackthorn, the larval foodplant of both black and brown hairstreaks, but various other species – and some trees – that attract or feed insect were included, such as oak, elm, field maple, buckthorn, dog-rose, privet and sallows, the food of the purple emperor butterfly. In addition, the old wood edge adjoining this land was left to grow unkempt, as was a hedge containing elm that separated the former arable and ley areas. The scraped exposures of Oxford clay between the banks were sown with wildflower seed mown from a nearby nature reserve.

Sowings and plantings occurred in 1990, with some replacement in 1991-92. Six years later, the shrubs and trees were well-established, including the *Prunus* which had suckered and begun to merge into genuine banks of scrub. This was quickly colonised by the brown hairstreak *Thecla betulae*, which today flies in abundance along the edges, egglaying up to the verge of the M40; its numbers rapidly increased in comparison to the population elsewhere in Bernwood Forest (Figure 5), and the compensation area currently supports a major population of this nationally scarce butterfly. The rarer black hairstreak *Strymonidia pruni* prefers more mature banks of *Prunus* and is a notoriously slow coloniser of new habitat (Thomas *et al.*, 1993). The site now appears suitable for the species, which first colonised the new habitat in 1999; this area is eventually likely to contain the largest area of optimum habitat for this species in the UK. The scarce white-letter hairstreak *Strymonidia w-album*, has also established a colony on elms *Ulmus procera* in the compensation area.

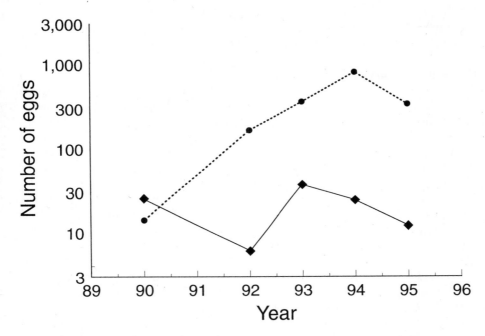

**Figure 5.**   Increase of Brown Hairstreak eggs on *Prunus spinosa* in the compensation area beside the M40 (dotted line) compared with average numbers on control hedges elsewhere in Bernwood Forest (solid line).

The sown grasslands have been equally successful. After the usual first years when species fluctuated between extremes of abundance and scarcity, an increasingly rich and stable sward is developing containing the characteristic (but now rare) flora of damp meadowlands on Oxford clays, including green-winged orchid *Orchis morio*, grass vetchling *Lathyrus nissolia*, bird's-foot trefoil *Lotus corniculatus*, pepper saxifrage *Silaum silaus*, cowslip *Primula veris*, hairy violet *Viola hirta*, yellow rattle *Rhinanthus minor* and quaking grass (Bickmore 1992, 1997, Chapter 24). In addition to the scarce hairstreaks, twenty-five species of (mainly grassland) butterfly have been recorded in the compensation area. Most are now established as self-contained breeding populations, which colonised and increased on the site in 1991-94; none is nationally rare, but three species – the brown argus *Aricia agestis*, marbled white *Melanargia galathea* and Essex skipper *Thymelicus lineola* are scarce in this region of England. Thus, although it may take 50 years for the full range of habitats and species contained in the design to develop, a diversity of valued plants and invertebrates is already supported by this formerly sterile area of intensive farmland, representing an object lesson on the potential for creating scrub and grassland habitats beside a busy motorway.

## Habitat creation on Twyford Down

It is beyond the scope of this paper adequately to describe the large-scale habitat creation that occurred following the construction of the M3 motorway through Twyford Down, east of Winchester. Much is summarised in a recent booklet (Snazell, 1998); more detailed results of the first 5-years' monitoring will be published in the scientific literature. However, in view of inaccuracies perpetuated at this conference, a résumé is desirable.

Following extensive Public Inquiries into various proposals to bypass Winchester, an easterly route was chosen in 1990 cutting through Twyford Down. As with the M40 near Oxford (and route B in Figure 1), the route ran primarily through arable land, avoiding all but a small fraction (c 0.15 ha) of the species-rich chalk grassland in the area; there was also concern that the road might isolate two colonies of chalkhill blue butterfly *Lysandra coridon*, breeding on The Dongas to the east and on St Catherine's Hill to the west. Due these concerns, it was decided to recreate 9 ha of chalk downland on former arable land on top of Twyford Down and along the route of the old Winchester bypass (A33), after its 19th century contours had been restored.

Preliminary surveys and historical researches were made to define the past and present fauna and flora of this site. Various types of habitat were then incorporated into the design, aimed at restoring the characteristic rich flora and invertebrate fauna that had been lost over the previous century from most of Twyford Down following agricultural 'improvement' and from parts of St Catherine's Hill when the A33 was constructed (Thomas *et al.*, 1991). Most of the enriched soil of the former arable land was removed and the rest mixed with bedrock to create a variety of calcareous seedbeds with varying degrees of (low) fertility on a range of aspects experiencing diverse microclimates. Similar seedbeds were prepared along some 2 km of the former A33 after a bedrock of chalk had been laid restoring its ancient contours. All (0.3 ha) of grassland of any conservation value on the motorway route was transported to these beds, either by hand or using a new technique of 'macroturfing' (Snazell, 1998), enabling 30 cm deep 2.4 m x 1.2 m turves to be transported and relaid either in solid blocks or in chequer boards with bare ground between, in areas large enough to eliminate damage from frost or drought, and thick enough (on chalk) to transport most soil organisms, including intact ant nests and relatively deep-rooted plants such as orchids. The available turf covered less than 4% of the compensation areas; the rest was seeded with various 'short' and 'medium-tall' turf mixes, chosen both to create a characteristic flora and to enhance the larval foodplants of the chalkhill blue *Lysandra coridon* and other downland butterflies. To collect the 260 kg of seed needed from local downland would have caused unacceptable damage, so commercial mixes were used containing seed of known provenance, virtually all species coming from south England and many from Hampshire; seed that was unavailable commercially was collected locally by hand and incorporated into the mixes. The seeding was supplemented by about 100, 000 plugs of key plants (including butterfly larval foodplants) that were unlikely to establish well by sowing. These were propagated from plants on St Catherine's Hill and other local downs, and included horseshoe vetch *Hippocrepis comosa*, kidney vetch *Anthyllis vulneraria*, rock rose *Helianthemum nummularium*, thyme *Thymus serpyllum*, cowslip *Primula veris*, hairy violet *Viola hirta*,, devil's-bit scabious *Succisa pratensis* and clustered bellflower *Campanula glomerata*. Two hundred juniper *Juniperus communis* bushes were also planted, increasing by fifty-fold the population of this declining shrub in this region.

The downland restoration occurred in three phases in 1992-94; after 1-2 years of protection from rabbits and other herbivores, the developing swards were managed by sheep grazing and occasional mowing. Unusually for a habitat restoration scheme, the fauna and flora are being monitored for the first 5-8 years. The first 3-5 years' results are extremely encouraging. The general appearance of all compensation areas is of attractive downland, rich in vetches, rockrose, thyme and – more locally – orchids in early summer, followed by an abundance of knapweeds and scabious in high summer. Indeed, anyone recalling the traffic jams of the old Winchester bypass would be enchanted now to walk the bridleway that has replaced it, with the Itchen canal on one side and downland containing an abundance of flowers and butterflies on the other. The vast majority of species in the seed mixes are now established in the downland areas, and the 58 'target' chalk grassland species of plant transported in turves in 1992 have now increased to 60 species (Snazell, 1998) including the frog orchid *Coelogossum viridae*. Invertebrates quickly colonised the site. All species of ant recorded on St Catherine's Hill and in the richest turf of The Dongas are well-established throughout the compensation areas. This is an important development, because ants are 'keystone' soil organisms which dominate the local environment at their scale, affecting the persistence and abundance of other species and the composition and biodiversity of the whole community. Most beetle, bug and spider species found on St Catherine's Hill and The Dongas have also colonised the new areas, although a few species are not expected until a deeper moss and litter layer has developed. Finally, the butterflies have been a particular success. Every species recorded on neighbouring downland now breeds on each of the three compensation areas, many in discrete colonies, and an additional species, the small blue *Cupido minimus*, which disappeared in this locality in the 1950s, has colonised all three areas. Furthermore, the chalkhill blue *Lysandra coridon*, the focus of the original protest, has thrived since the opening of the M3. The former colonies on St Catherine's Hill and The Dongas have increased slightly in size since the opening of the M3, and have been joined by a third colony of several hundred adults, breeding on the oldest of the three compensation areas. Mark-recapture experiments in 1996 established that not only was this a new self-contained population, but that a small number of adults flew between the three sites, including across the motorway, indicating that the metapopulation structure for this species east of Winchester has been improved by the compensation areas rather than been damaged by the changed traffic route resulting from the construction of the M3.

## CONCLUSIONS

The results of research on insect behaviour and populations, and of recent attempts to recreate valued habitats, suggest that new roads provide considerable  opportunities to enhance rather than harm the populations of common and scarce species in a locality, provided the route crosses impoverished ground for wildlife (e.g. intensive farmland). The success of habitat creation beside the M40, M3 and elsewhere should not be used to justify fixing routes across rich semi-natural biotopes because, with current knowledge, this will almost certainly result in a net loss of wildlife. Some terrestrial habitats, notably wetlands and ancient woodland, are particularly difficult or impossible to establish, at least within decades rather than centuries; heathland, calcareous and neutral grassland are much simpler. But even with these biotopes there is a need for further research both on the habitat requirements of key invertebrates and on improving techniques to create and maintain them. Particular gaps in knowledge include how to create a characteristic soil fauna on new ground, and how to design verges that act as conduits for the dispersal of insects.

# REFERENCES

**Anon. 1987.** *Tagfalter und ihr Lebensraum.* Geiger W. ed. Schweizerisches Bund fur Naturschutz, Basel. 354-368.

**Bickmore CJ. 1992.** M40 Waterstock-Wendlebury: planning, protection and provision for wildlife. *Proc. Instn Civ. Engrs Mun. Engr.* **93:** 75-83.

**Bickmore CJ. 1997.** *M40 waterstock to Wendlebury: monitoring report.* London: Symonds Travers Morgan.

**Dennis RLH. 1986.** Motorways and cross-movements. *Bulletin of the Amature Entomological Society.* **45:** 228-243.

**Dover JW. 1991.** The conservation of insects on arable farmland. In: Collins NM, Thomas JA. eds. *The conservation of insects and their habitats* London: Academic Press. 294-318.

**Feltwell J, Phillip J. 1980.** The natural history of the M40 motorway. *Transactions of the Kent Field Club.* **8:** 101-15.

**Free JB, Gennard D, Stevenson JH, Williams IH. 1975.** Beneficial insects present on a motorway verge. *Biological Conservation.* **8:** 61-72.

**Pollard E, Elias DO, Skelton MJ, Thomas JA. 1975.** A method for assessing the abundance of butterflies in Monks Wood National Nature reserve in 1973. *Entomologist's Gazette.* **26:** 79-88.

**Mader HJ. 1984.** Animal habitat isolation by roads and agricultural fields *Biological Conservation.* **29:** 81-96.

**Morris MG, Thomas JA, Ward, Snazell RG, Pywell RF, Stevenson M, Webb NR. 1994.** Recreation of early successional stages for threatened butterflies - an ecological engineering example. *Journal of Environmental Management.* **42:** 119-135.

**Morton ACG. 1985.** The population biology of an insect with a restricted distribution: *Cupido minimus* Fuessly (Lepidoptera, lycaenidae). [PhD thesis], University of Southampton.

**Munguira ML, Thomas JA. 1992.** The use of road verges by butterfly and burnet populations, and the effect of roads on adult dispersal and mortality. *Journal of Applied Ecology.* **29:** 316-329.

**Port GR, Spencer HJ. 1987.** Effects of roadside conditions on some Auchenorryncha. *Proceedings of the sixth Auchenorrhyncha Meeting* Toronto. 7-11.

**Port GR, Thompson JR. 1980.** Outbreaks of insect herbivores on plants along motorways in the United Kingdom. *Journal of Applied Ecology.* **17:** 649-56.

**Ravenscroft NOM. 1994.** The ecology of the chequered skipper butterfly *Carterocephalus palaemon* Pallas in Scotland. II Foodplant quality and population range. *Journal of Applied Ecology.* **31:** 623-630.

**Snazell RG. 1998.** *Ecology and Twyford Down.* Wareham: Institute of Terrestrial Ecology.

**Sutcliffe OD, Thomas CD. 1995.** Open corridors appear to facilitate dispersal by ringlet butterflies (*Aphantopus hyperantus*) between woodland clearings. *Conservation Biology.* **10:** 1359-65.

**Thomas CD, Thomas JA, Warren MS. 1992.** Distributions of occupied and vacant butterfly habitats in fragmented landscapes. *Oecologia.* **92:** 563-567.

**Thomas JA. 1983.** A 'WATCH' census of common British butterflies. *Journal of Biological Education.* **17:** 333-338.

**Thomas JA. 1984.** The conservation of butterflies in temperate countries: past efforts and lessons for the future. In: Vane-Wright RI, Ackery P eds. *Biology of Butterflies, Symposia of the Royal Entomological Society 11.* London: Academic Press. 333-353.

**Thomas JA. 1986.** *RSNC Guide to Butterflies of the British Isles.* London: Country Life Books (Hamlyn).

**Thomas JA. 1991.** Rare species conservation: case studies of European butterflies. In: Spellerberg I, Goldsmith B, Morris MG. eds. *BES The scientific management of temperate communities for conservation. Symposium 29.* Oxford: Blackwells. 149-197.

**Thomas JA, Lewington R. 1991.** *The butterflies of Britain and Ireland.* London: Dorling Kindersley.

**Thomas JA, Moss D, Pollard E. 1994.** Increased fluctuations of butterfly populations towards the northern edges of species' ranges. *Ecography* **17:** 215-220.

**Thomas JA, Munguira ML. 1991.** Effects of roads on butterfly and burnet populations. Abbotts Ripton: Annual Report Institute of Terrestrial Ecology 1989/90. 53-55.

**Thomas JA, Ward LK, Snazell RG, Pywell R. 1992.** *M3 Suggestions for the restoration/re-vegetation of the motorway banks and the compensation area at the Arethusa Clump.* Wareham: Institute of Terrestrial Ecology.

**Wade KJ, Flanagan JT, Currie A, Curtis DJ. 1980.** Roadside gradients of lead and zinc concentrations in surface-dwelling invertebrates. *Environmental Pollution Series B* **1:** 87-93.

**Warren MS. 1987.** The ecology and conservation of the heath fritillary butterfly *Mellicta athalia* II. adult population structure and mobility. *J. Appl. Ecol.* **24:** 483-98.

**Wells TCE, Bayfield NG.** *Wildflower swards for trunk roads and motorway landscaping: a handbook.* Abbotts Ripton: Institute of Terrestrial Ecology.

# 20

# Measures to protect
# amphibians and reptiles from road traffic

A E S LANGTON

Setting Quality Standards in mitigation: system design, construction, maintenance and monitoring.

Measures to protect vertebrates from vehicle collision, in order to protect both animals and vehicle passengers, have largely been developed during the second half of the 20th century. Most obvious are signs at the side of roads warning of crossing animals. More recently fence and tunnel systems have been built to reduce or prevent the likelihood of collisions.

Britain is a small temperate island with a depleted vertebrate fauna, and measures have been largely confined to systems to reduce collisions between vehicles and deer, badger *Meles meles*, otter *Lutra lutra*, and common toad *Bufo bufo*, with interest noticeably increasing since the mid 1970s (see Barton, this volume). Measures include the use of specialised road warning signs, fences and tunnels/overbridges. These are sometimes built during road construction as a result of the recommendations of Environmental Impact Assessments (EIAs) or other ecological appraisals.

In mainland Europe, the variety of vertebrates increases greatly and in some cases efforts to prevent vehicle/animal collision go back to the middle of the century and involve larger and heavier animals such as elk *Alces alces* and reindeer *Rangifer tarundus,* as well as rarer species such as wolf *Canis lupus* and lynx *Lynx lynx*, in Finland (Luukkainen, 1989).

Worldwide, tunnel systems that have been built into road and rail transport networks have also been the subject of study. One example is the use of culverts and tunnels (up to 3 metres in diameter) by small mammals such as long-nosed bandicoot *Peromelus nasuta*, and swamp wallaby *Wallabia bicolor*, in open forest and sub-tropical rainforest of New South Wales, Australia (Hunt *et al.*,1987).

Outside Europe however, other than for road signs and fencing, there are few examples of species specific mitigation measures in the published literature but interest is growing (Trombulak & Frissell, 2000). Road bridges that span valleys and rivers or that are otherwise raised up (e.g. viaducts) allow relatively unhindered movement of most animals below them. Roads may built underground or in tunnels to allow large animal movement above them, for example the Hatfield Tunnel on the M25 motorway in Hertfordshire, England. Ecoducts or wildlife overpasses (Berris, 1997) are bridges (up to 200 m wide) that are lined with soil and natural vegetation to encourage wildlife use. These have been used in the Netherlands, Germany, Switzerland, USA and Canada and may enable all sizes of animals to cross roads. Bridges and tunnels (or culverts – see Yanes et al 1995) and underpasses (e.g. see Foster & Humphrey, 1995) have been constructed in the USA and Canada, principally for use by larger, often endangered species, including of bear *Ursus* sp., deer *Odocoilus* spp., panther *Felis concolor*, wolf *Canis lupus,* and goat *Oreamnos ameicanus* (see review in Jackson & Griffin, 1998). The impact of animal/vehicle collision (e.g. for mammals see Oxley *et al.,* 1974) and interest in mitigating it in central and northern Europe has resulted in the installation of hundreds of tunnel systems, mainly for badgers and amphibians in Germany and Switzerland and smaller numbers of systems in Austria, the Netherlands, France, Spain, England, Hungary and Belgium. However, there appears to have been relatively little activity elsewhere in the world.

## AMPHIBIANS

Measures taken for amphibians exceed those for reptiles, partly because their synchronised seasonal movements are (with a few exceptions such as some snakes and chelonians) more noticeable than those of reptiles.

In the UK, a standard warning sign approved by the Department of Transport (No.WBM (R)551.1) in 1985 has been largely confined to England. It depicts a common toad *Bufo bufo* although its use may benefit other amphibians, including the frog *Rana temporaria* and the three newt species *Triturus cristatus*, *Triturus vulgaris* and *Triturus helveticus*. Since the first signs were tried out in 1984, approximately 600 sites in England and Wales have been registered as suitable for signs, (Figure 1.) These are places where hundreds or thousands of adult toads may be killed each year during spring movements/migration to and from their spawning waters, and emerging toadlets may also be killed upon their dispersal in early summer. In England, the first tunnel and fence system was installed in 1987 (Langton, 1989a) and since then at least fifteen tunnel and fence systems have been built.

Continental Europe has many more species of frogs, toads and newts than Britain and in addition salamanders. Road warning signs (and patrols to

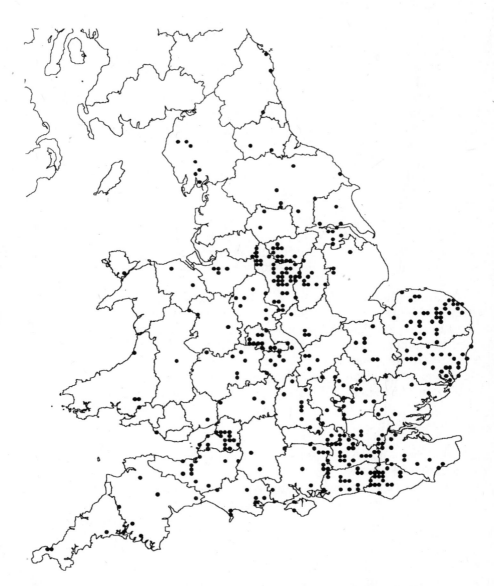

**Figure 1.** 600 sites in England and Wales have been registered as suitable for signs.

rescue amphibians) have been used most consistently and in a co-ordinated way in Switzerland (Grossenbacher, 1981), Germany (e.g. Anon, 1992: Stolz & Podloucky, 1983) and the Netherlands (van Leeuwen, 1982) with designs also used in Austria, France, Italy, Belgium and a few other adjoining countries.  Ad-hoc amphibian sign designs, sometimes produced by local governments but more often by concerned local organisations and individuals also appear sporadically on public and private roads elsewhere in the world.

Tunnel and fence systems have been extensively used in Germany and Switzerland where such measures are sometimes a legal requirement for mitigating the damaging effects of road building (Langton, 1989b). Systems also exist in adjacent countries and, although no definitive record appears to be kept in any country, it is clear that there are hundreds of systems in Germany and Switzerland (possibly in excess of 500 systems) while the number in Belgium, France, England, Spain, Italy, Austria, Hungary, Czech republic, Poland, Norway, Sweden and Denmark is likely to be less than 100 in total (T. Langton, *pers. obs.*). This nevertheless represents a significant investment, estimated from the known range of costs of system installation in different countries at between 15 - 30 million (1998 prices) in public funds over the last thirty years, with the earliest systems dating back to the 1960s.

Outside Europe there seem to be few examples of measures to specifically protect amphibians crossing roads, although several systems exist to accommodate all, or most, wildlife species. Examples of tunnels for single amphibian species are the first salamander tunnel installed in the USA (Jackson, 1996: Jackson & Tyning, 1989) and the tunnels reported for salamanders in Japan.  New designs and prototype systems are currently being considered in the USA (Jackson, 1997: Jackson & Marchand, 1998).

## REPTILES

In Britain death on roads is a reported cause of decline in snake populations. However, most lizards appear to attempt to cross surfaced (paved) roads less frequently, perhaps as a function of their having a smaller home range.  Chameleons are killed in significant numbers crossing some roads in southern Spain (CODA 1992). In tropical central America, Jesus Lizards *Basilicus* spp., are frequent casualties (J. Burton, *pers. com.*). In Europe, snake, tortoise and terrapin populations may be impacted, although most species may benefit from the bridge systems that span streams, rivers and wetlands. Tortoises, lizards and snakes may benefit from the large supported bridges crossing drier valleys. Vehicle traffic is considered one of the greatest threats to the rare and threatened Cyclades blunt-nosed (Milos) *Viper macrovipera* schweizeri (C.Andren, G. Nilson *pers. com.*). There are a few examples of purpose-built reptile tunnels. The first European tortoise tunnel in France was constructed for Testudo hermanni in the Department of Var as part of the A57 Autoroute in 1991 (Devaux, 1992). There have been a number of fence and tunnel (culvert) systems created specifically for tortoises in the desert habitats of S W USA (Boarman & Sazaki, 1996).

The crossing of, or basking on, paved roads by reptiles may particularly occur when roads are built through pristine or relatively undisturbed areas. Reptiles are also common in modified habitats where drainage lines or wetlands are crossed by roads, and freshwater turtles (e.g. Foule, 1998) and crocodilians may be killed in large numbers. Nocturnal reptiles (especially snakes) make use of roads and road edges to gain the residual heat that is absorbed and slowly released by road surfacing material. Examples include the rattlesnakes *Crotalus* spp. in North America that range widely, crossing and basking on roads, and tropical boas and pythons that are frequently killed on roads. Red-sided garter snakes are similarly at risk when they den close to main roads in Manitoba, Canada. Rocks and boulders heaped up to build or protect roads may often provide refuges for reptiles and their prey at the sides of roads.

Road signs warning of tortoises, crocodilians and snakes crossing roads or tracks in protected areas and other places are perhaps the most widespread anti-collision measure but are still confined to relatively few countries. As with warning signs for all forms of wildlife there may be great regional variation in the design and extent of their use, and little or no advice is given to local people on how to minimise the risk of vehicle collision with reptiles and other wildlife.

## STANDARDS

The rapid growth in use of motor vehicles, increasing spread of road networks and increasing vehicle traffic density and speed in recent decades has created a wide range of variables that make the study of their impact on the environment hard to measure. Coupled with their direct impact, new road construction often facilitates human access to the near-road environment, enabling secondary and more intensive land-use changes to take place such as building construction and human occupation, forest logging and agriculture. Such secondary changes may make the study of the impact of the road itself on wildlife difficult or impossible to gauge in isolation. The less obvious and cumulative effects such as that from pollution (see Ward this volume), adds to the difficulty in isolating the effects of single factors.

There appears to be little standardisation in the approach to designing wildlife mitigation measures for new or existing roads. Factors that influence the use of tunnels and bridges as 'wildlife passage structures' by animals (Jackson & Griffin, 1998) are often species specific. Despite a limited amount of study it is clear that for mammals and some reptiles at least, the degree of acceptance and rejection of tunnel/underpass/overbridge systems varies with behavioural needs/sensitivities of species. Some design requirements for different species may in fact conflict with each other. There is little of no evidence that mitigating measures can retain exactly the same pattern of movement (population fluidity) in species living in a divided wildlife community than was present before a road was built. Mitigation measures are therefore aimed at minimising change, rather than seeking to eliminate it, which may be impossible for many species.

One strategy aimed at creating a consistent approach to building physical structures to enable wildlife movement across roads is based on a 'landscape analysis' to identify 'connectivity zones' (Jackson & Griffin. *op. cit.*). This requires the broad based evaluation of wildlife distribution and movements along a road route. Mitigating installations are generally designed where roads cross streams, rivers or wetlands, and are also built where ecological studies indicate important species/habitats corridors that will require the provision of wildlife overpasses or bridges. Some studies of small animals concern measurements of them using culverts installed for water drainage as opposed to wildlife passage purposes. The first examples of system quality standards for purpose-built systems appear to be in Germany and Switzerland. In some German Landers in the mid 1990s, for example, the use of 200 mm diameter tunnels for amphibians was lapsed in favour of 500mm tunnels, as a minimum recommended diameter (Podloucky, *pers. com.*). Distance between tunnels is also regulated during Statutory (mandatory) consultation between road builders and government nature conservation authorities. Elsewhere, specification of systems is often done on an experimental basis, and decisions on the type of material, tunnel size and position reflects the limitations of incorporating such measures into a previously approved engineering design. This may compromise the effectiveness of the system.

## AMPHIBIAN TUNNELS IN ENGLAND

The evaluation of tunnel and fence systems that have been installed in England for amphibians has been undertaken in order to compare their performance against design characteristics and to help the consideration of ways to improve the chances of future installations being successful.

## Methods

Tunnel and fence systems that have been installed to mitigate the effect of road construction on amphibian populations in England (1987-1997) were reviewed by gathering information from those involved in system design, installation and monitoring.   A quantitative scoring system for the assessment of the following factors was developed; i) state of completion of each system, ii) adequacy of repairs and maintenance of tunnel and fence systems post-construction, and iii) the monitoring of the extent of use of the systems by amphibians.  The maximum number of points allocated in each category was 3 (range from 0-3) with 3 points indicating the highest level of achievement.

### *State of completion*
Each system was scored up to three points.  One was allocated if the system designer considered that the appropriate number of tunnels had been built and placed in the correct position along the road route. A second point was allocated if the appropriate length of guidance fencing had been correctly placed either side of a road and in the correct position to direct amphibians towards the tunnel entrances. A third point was allocated if tunnel entrances or other devices were constructed at tunnel entrance/exit positions in order to guide amphibians into the tunnel entrances and to prevent them walking past the entrance/exit positions.

### *Adequacy of repairs and maintenance*
One point was allocated if the system had been repaired following damage caused by vehicle collision, damage from objects thrown or falling from vehicles, vandalism, accidental damage or faulty installation. One point was allocated if the adopting authority had carried out annual (twice yearly) cutting of grass along the fence line. A third point was allocated if the inside of tunnels had been checked each year to remove blockages, repair or replace soil substrate at the base of tunnels and if necessary the jetting out of accumulated petroleum/rubberoid residues on the tunnel floor (surface tunnels ACO type).

### *Monitoring of use of by amphibians*
Over the working life of each tunnel since its installation, a judgement was made as to the level of monitoring that had been carried out at each site.

**Figure 2.** Location of fifteen amphibian tunnel systems in England documented for this review.

One point was allocated if at least some level of interest had been shown with ad hoc observations by at least one individual each year. A second point was allocated where every year, or for at least three years since construction, checks had been made on tunnel use through visits spread throughout spring and early summer (March - July). A further point was allocated if every year, or for at least three years since construction, detailed daily measurements using pitfall traps at the end of each tunnel (or other monitoring system) had been made in order to determine level of use of the system by amphibians.

## Results

Information was collected from fifteen sites around the UK (Figure 2) where amphibian tunnel systems have been installed. All of the systems are in England. Limited information was available for a further two systems but there was insufficient detail for their inclusion in this analysis. It is considered unlikely that more than 17 purpose-built amphibian tunnel systems exist in the UK in 1998. Of the fifteen tunnel systems (Table 1), nine were installed during of the construction of new roads, five were the result of installing systems into existing roads and one was on private land.

Most of the systems were installed on busy 'A' grade roads, these being two or four lane paved roads with speed limits of 30 - 60 miles per hour (48 - 96 km/hr). Most systems had two tunnels (range 1 - 6 tunnels). A total of 32 tunnels had been installed. Eleven of the fifteen sites (23 or 71% of tunnels) were constructed with ACO surface tunnels of 200 - 500 mm width (see figure 3a,b), the other sites comprising concrete pipe (3 sites, 5 tunnels), ceramic pipe (1 site, 3 tunnels) and a 1000 mm wide concrete shelf built into a river culvert (one site). One site had two types of tunnel installed.

Tunnel lengths varied and this reflected both the width of the road and the width of verge at each side. Maximum tunnel lengths ranged from 40 metres, where tunnels crossed under dual carriageways or motorway grade roads, down to about 4 metres for tunnels crossing roads with two way, single-lane traffic. Most tunnels were within the 14 - 20 metre length category.

Permanent amphibian fences were placed on one side of a road (6 sites) or opposite each other on both sides of a road (7 sites) or on multiple sides where road junctions were involved (1 site). One site had no fence. Fence lengths varied from short headwalls a few metres wide, to those 1000 metres in length, with six sites having fences of 100 - 175 metres either side of tunnel entrance/exit points. At least nine sites ACO Wildlife (cast

**Figure 3.** (a) An ACO precast tunnel. (b) The tunnel *in situ* on a new road.

**Table 1.** Information collated on the 15 amphibian tunnel use in England 1987-1998.

| Year instlld | County Grid Ref, | Road No. | No. & type tunnel(s) | Tunnel length | No. & length Fence | Amphibians Target (bold) other spp. | Other species using tunnel. |
|---|---|---|---|---|---|---|---|
| **1987** Tunnel Fence | Bucks SU 76 85 | A 4155 | 2: Q200 ACO | 2 @ 10 | 1(nth) 300 ACO | ***Bufo bufo*** | Ducklings and small mammal |
| **1988** Tunnel | N.Yorks SE 71 55 | | 2:Q200 ACO | 2(T) temp | Data incomplete | ***Bufo bufo*** | Data incomplete |
| **1989** Tunnel Fence | Norfolk TL 85 83 | A11 | 3: 2 Q200 1 500mm pipe | 2 @ 10 1 @20 | 1 (wst) 200 | ***Bufo bufo*** | *Natrix natrix* |
| **1989** Tunnel | Cambs TL 20 95 | Buntings Lane | 2: 300 Concrete pipe | 2 @ 15 | 2 (Head wall only Both sides) 15 | ***T. cristatus*** *Bufo bufo* *R. temporaria* *T. vulgaris* | Snail spp. |
| **1990** Tunnel | S. Yorks SE 29 00 | Cote Lane off A629 | 1: Q200 ACO | 1 @ 15 | None | ***Bufo bufo*** | None |
| **1990** Tunnel **1992** Fence | Sussex TQ 16 12 | A283 | 2: Q200 ACO | 2 @ 10 | 1 (nth side) | ***Bufo bufo*** *R. temporaria* | None |
| **1992** Shelf Fence | Norfolk TG 12 00 | A11 | On river bridge 1000mm | c. 40 | 2 (both sides) 3000 ACO | ***T. cristatus*** *Bufo bufo* *R. temporaria* *T. vulgaris* | (River) |
| **1993** Tunnel Fence | Lancs SK 40 17 | A512 | 3: 500mm ceramic pipes | 1 @ 17 1 @ 19 1 @ 22 | 2 (both sides) 1200 paving slabs | ***Bufo bufo*** *R. temporaria* *T.vulgaris* | *Meles meles* |
| **1995** Tunnel Fence | Sussex SU 97 22 | A283 | 1: AT200 ACO | 1 @ 8 | 1 (east) only 30 on west 305 ACO | ***Bufo bufo*** *R. temporaria* *T.vulgaris* | None |

| 1995 Tunnel Fence | Cheshire SJ 84 83 | A34 | 2: 600mm Concrete pipes with turf | 2 @ 35 | 2 (both sides) 980 ACO | *Bufo bufo* *T. cristatus* R. temporaria T. vulgaris T. helveticus | Data incomplete |
|---|---|---|---|---|---|---|---|
| 1995 Tunnel Fence | Suffolk TM 13 78 | A143 | 2: AT500 | 2 @ 15 | 2 (both sides) 200+ACO 2 grids | *Bufo bufo* T. critatus | *Natrix natrix* |
| 1995 (2) Tunnel Fence | Cambs TL 39 60 | B | 1: AT200 ACO | 1 @ 13 | 1 (both sides) 30 ACO (460T) | *Bufo bufo* R. temporaria T. critatus T. vulgaris | None |
| 1996 Tunnel Fence | Notts SK 48 43 | A 6096 | 6: AT500 ACO | 6 @ 20 (30 m apart) | 2 (both sides) 300 ACO | *Bufo bufo* R. temporaria | R. ratus N. natrix |
| 1996 Tunnel Fence | Bucks SU 93 80 | Prvt off M4 slip | 2: AT500 ACO | 2 @ 4 | 1 700 | *T. critatus* R. temporaria | No data |
| 1998 Tunnel Fence | Glos SO 39 21 | A40 | 2: AT200 ACO | 2 @ 14 | 4 (both sides) 40 | *Bufo bufo* | None |

recycled plastic) fencing had been used, this being the only purpose built fence that has been marketed in the UK. At other sites concrete wall or paving slabs had been used to form a barrier to amphibian movement.

The species for which most systems had been designed was the common toad *Bufo bufo* at ten sites (figure 4), and without this species being present it is considered unlikely that the system would have been built. At five sites, the legally protected great crested newt *Triturus cristatus* was present, and at three of these the presence of the species and its listing under the Wildlife and Countryside Act (1981) (and other more recent legislation) had created a legal requirement for measures to be taken to protect the species from road construction and traffic. At three sites the great crested newt was one of a four or five species amphibian community of local or regional significance.

At one site, the movement of grass snakes (*Natrix natrix*) from their breeding habitat (farmyard) to feeding/basking area (newly constructed lake in river valley) had provided additional justification for fence and tunnel

**Figure 4.** Toads at the entrance to an ACO tunnel.

measures. A few other species such as badgers and a range of other animals were also reported as using tunnels to some extent.

The results of the assessment of tunnel installation quality are shown in figure 5. Only five systems (33%) were considered to have been installed to a full design specification, while most were incomplete to a greater or lesser extent. Level of maintenance varied, with most sites receiving at least some maintenance, but only four sites were monitored in a consistent way. Even at these sites monitoring was confined to occasional checks over one or two years, and usually only during the amphibian breeding season following the year of installation. Monitoring was most frequently carried out by private individuals, sometimes on behalf of a local nature conservation group but there was official monitoring by local government staff/consultants and/or University staff/students at four sites.

At seven sites the systems had been installed primarily as the result of campaigns by private individuals, local charities and other organisations, while at seven others the measures were a part of planned mitigation for new road building by the Department of Transport (now DETR) and County Council Engineering Services. At one site the installation had been part of a condition placed by the local Planning Authority for construction of a new housing scheme. In all but two schemes, the responsibility for adopting and maintaining the tunnel and fence systems is (or will be) that of the local County or District Authority. However, the responsibility for fencing installed beyond the official highway (verge) boundary with private land, may only be partly adopted for maintenance and repairs if a specific undertaking has been made to allow it.

## DISCUSSIONS

### General

The level of interest in mitigating the effect of roads on nature conservation has grown considerably in Europe over the last three decades, and some countries are developing better guidance on mitigation impacts (e.g. for England see Anderson 1993). The problem of direct loss of wildlife habitat and the need for sympathetic landscaping of road embankments and cuttings using local materials has received considerable attention, (e.g. for Scotland see Scottish Office 1998). Most attention to environmental damage has addressed issues of impact on water systems (including changes to catchment systems and the effects of contaminated run-off), air pollution and the impacts of noise and light on people. Although many of these issues also have direct bearing on wildlife, the longer-term issue of how animal (and plant) communities are influenced by reduced dispersal capabilities caused by new road construction has emerged relatively slowly as a cause for concern.

Road warning signs appear to be effective in alerting motorists, to reduce their vehicle speeds and helping to reduce animal/vehicle collisions. They may also help to protect the increasing numbers of volunteers rescuing small animals. In recent years two amphibian rescue volunteers (one in England and one in Belgium) were injured by passing vehicles. More consistent use of warning signs and better guidance for motorists on their meaning is required in most, if not every country. Cost of sign manufacture, placement and the theft of signs for their scrap metal value has been a significant factor in the extent of their use in some countries.

(a)

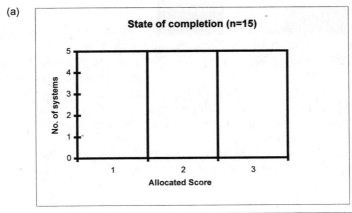

(b)

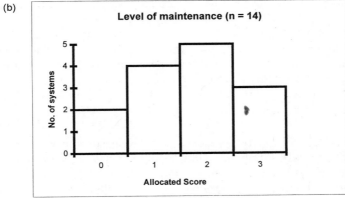

(c)

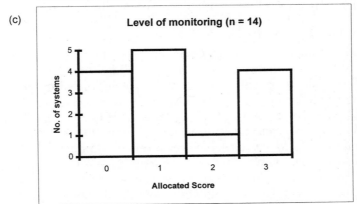

**Figure 5 (a), (b), (c).** Allocated scores (see text) for the completion level of maintenance and monitoring of amphibian tunnel and fence systems documented in this review.

The use of fencing to control movements of animals requires careful attention as small animals may get stuck and die from starvation or exposure/overheating when snared in wire or plastic fence of small diameter (10 - 50mm) and birds may be killed or injured flying into taller fencing.

Many, perhaps most new animal underpasses/overbridges must be considered as experimental. Judging the success or failure of systems is difficult and likely to require long term monitoring using well thought out criteria.

In general there are widely varying levels of interest on the impacts on roads on wildlife in different parts of the world. Many developing countries may be subject to socio-economic constraints that would give a low priority to promoting mitigation measures. In some instances the impact of traffic is dependant on its volume (number of vehicles per hour) and duration, and the need for measures is likely to be highly localised. There is a need to collate information on such localised need in each country in order to design a balanced strategic approach to promoting a realistic level of mitigation. This may benefit from international co-operation to share experiences and technology that is rapidly developing (see later).

**Amphibians and roads**

In Britain, traditional approaches to the issue of animals crossing roads have been relatively simplistic. Until recently it has been considered that the conversion of road embankments to semi-natural habitat, giving often a net gain in such habitat type, is adequate compensation for any negative effect of road building on wildlife. While some species, in addition to plants, such as small mammals, butterflies, some birds and reptile species may potentially increase their local distribution as a result of verge creation, the wider and long term impact of roads on wildlife is less well documented.

The suggestion that there are "no indications that kills of individual animals on roads are in any way significant when related to the population of the species in the area" (May, 1970) now requires reconsideration, and the statement "It can be argued that it is better to have two animals killed on the road and eight others surviving in the habitat along the road than no habitat and no animals" (May, op .cit.), requires careful interpretation.

Published quantitative studies of amphibian mortality on roads have been relatively scarce. Preliminary approaches (van Gelder, 1973) do not appear to have been followed up with more than a few detailed studies. A four year study (1978-81) of common toad *Bufo bufo* migration to a small lake in mid-Wales (Gittins *et al.,* 1980: Gittins, 1983a), concerned a breeding site with a low level of traffic moving around the lake, mainly to gain access to a golf club. In one year of study the diversion of a small amount of local traffic increased adult mortality on that road from very low numbers to about 4% (Slater *pers. com.*). Male toads tend to sit on roads that immediately surround ponds watching for approaching females. The road mortality in the mid-Wales study was measured at low hundreds, for a population estimated to be in the region of 6,000 adult toads. The estimated natural mortality rate of adult toads each year at this (and another nearby) population however averaged about 60%. When compared with the 4% road kill mortality and publicised (Gittins, 1983b), it was suggested that, with respect to toads being killed by traffic in general "the overall impact may be insignificant". However it was not emphasised that the study site was untypical, having a low mortality of just 4% and for a single year. However the conclusion that the local extinction of toads due to road traffic is rare can be found even in relatively recent texts (e.g. see Beebee, 1996). In the mid-Wales lake study, there was a 3:1 adult male-female ratio, and very low female mortality was caused by traffic. For this large population, loss of emerging toadlets to traffic was also a tiny fraction of mortality. It was in fact the high natural mortality estimates of adult toads that were the basis for suggesting that, in general mortality rates from vehicles can be high without creating an effect. In another study Gittins (1983c) measured 'in-pond' mortality rates of females at about twice that of males. If this was partly a function of sampling technique ("as some toads evaded capture leaving the pond") then the 28% 'in-pond' annual mortality of toads measured may be an overestimate.

For a combined 'land and water' mortality rate of 60% per year, with toads living up to eight years, age-specific mortality may be a significant factor, as death from being run over may be more random that the factors causing the death of inexperienced young animals. Fitness/selection processes may be significantly disrupted.

It is noteworthy that ten years after the study at the mid-Wales lake the toad population has declined to very low levels, perhaps as a result of water quality changes, not from road traffic (Slater *pers. com.*). This reinforces the possibility that during the mid-Wales lake study the high mortality rates for the 1978-82 years reflected studies on a population that was close to its maximum carrying capacity. While other unpublished studies suggest lower adult mortality rates may be critical to population survival, it is likely that the point at which amphibian mortality from vehicles on roads begins to influence breeding success and recruitment and threaten populations will vary between species and according to the range of variables at each site. Such variables might include the proximity of a road to a breeding site, the amphibian population size and the proximity of alternative breeding sites which might enable natural re-colonisation of depleted areas.

The effect on toad populations of mortality of toads on roads is likely to vary according to population size, vulnerability of each sex/age class to mortality and other factors. For common toads *Bufo bufo*, a precautionary approach might justify concern when mortality exceeds 50% of natural 'in-pond' mortality. Notification of UK toad migration sites as suitable for the placement of warning signs showing toad migration routes across roads normally takes place where it appears that mortality on roads is approaching or exceeding 10 - 15% of the estimated adult population (usually hundreds of toads, as populations of toads do not often exceed 6,000 adults). Often, road traffic mortality will be only one of a number of factors operating with a negative impact on the population. The most important point is that animal mortality on roads may not need to be high to have a significant effect, when other pressures are also acting.

More recent studies have documented the decline of a large toad population where road traffic density adjoining the site has greatly increased but other factors such as land-use change, may also have influenced toad survival. One questionnaire study of 76 monitored toad populations in England where road traffic kill toads each year (Froglife, 1996) indicated that 75% of populations had undergone slight or substantial decline within the previous decade. However, population size is likely to be changing over such a timescale and interpretation must take this into account. In that study however, increase in road traffic was the biggest factor reported for the decline in toad numbers at the sites. Fahrig *et al.* (1995) found a significant negative effect of traffic mortality on the local density of frogs and toads in two regions near Ottawa in Canada.

A further complicating factor to the consideration of road/vehicle impacts on amphibian populations (especially in urban/suburban areas) is the effect of habitat fragmentation and reduced gene-flow on the survival of populations. Added to this are the knock-on effects of any change in numbers of a single species to the wildlife community as a whole. The loss of genetic variation and a change in species diversity and richness may have dramatic long term implications. This is an area of investigation that is rapidly developing due to the advancement of molecular techniques to examine and compare genetic variability within and between populations.

Investigations into the influence of roads on the genetic structure of populations of the common frog *Rana temporaria* in south-west Germany (Reh, 1989: Reh & Seitz, 1990) have indicated that in one study area, frog populations had become genetically impoverished due to isolating effects of motorways and railway lines (see also Hels, 1999). Evidence is now increasing for the need to take a precautionary approach to the possible long-term influence of roads and road traffic so that measures are incorporated with some degree of safety margin.

## Amphibian tunnels in England

The results of the review of fifteen amphibian tunnel systems in England suggest that one or more new systems are installed each year. About half are the result of statutory requirements and half in response to the pressure from local people to protect a local site where amphibian mortality (and other problems) have been evident.

The ACO surface drainage and wildlife fence system is the only tunnel and fence system marketed specifically for wildlife use in England. Tunnel and fence design specification may be critical to the success of a system (Dexel, 1989). The difficulties that amphibians encounter due to the dry caustic nature of alkaline residues in cast concrete pipes is well known. However, larger buried concrete pipes with a wide (1000 mm+) base made up of soil and vegetation have become more popular in recent years in some German Landers (Podloucky *pers. com.*).

Shortfalls in the completeness of the English amphibian tunnel systems gives cause for concern. One of the main problems appears to be the limitation of installing systems within the available land that has been acquired for road building; land that may be purchased well in advance of any detailed mitigation studies. Factors such as the correct positioning of fencing to channel amphibians towards tunnel entrances may be compromised. Development of new tunnel and fence materials and their use results in what is still, in practice, a rather experimental approach. Perhaps the greatest concern however is the low level or complete lack of detailed monitoring of the use of systems by amphibians and other animals at many sites. Because the overall efficiency of tunnel systems is not fully known, expectations for systems performance has been uncertain. In many places the precautionary approach of constructing substitute amphibian breeding ponds in the terrestrial habitat separated from the breeding pond by road building has been encouraged (Schlupp & Podloucky, 1994), as a safeguard to maintaining a large amphibian population size in the general area around a road.

Even with optimal alignment of fencing to guide amphibians to tunnels and adequate number and design of tunnels to allow their passage and return, one important though unquantified factor seems to be the proportion of amphibian (and other species) that use tunnels. Following the year of tunnel and fence installation, surviving adults with experience of finding a breeding pond on the other side of a new road may search to find it and this assists the process of exploring a new tunnel system. However, unless young amphibians moving from a breeding pond for the first time find and use a fence and tunnel system, the reduced terrestrial dispersion resulting may produce an isolating effect from the road. This influence can be reduced with the assistance of substitute ponds. The aim of future systems in some cases may well be to maintain two new amphibian breeding ponds on either side of the road, and close to the tunnel entrances with a level of interchange through tunnels aimed primarily at maintaining adequate gene flow. The role of the substitute pond is then more as a 'supplement' than a substitute and may become an essential component of all systems in the future. They can also be incorporated into systems already installed. The number of ponds needed will therefore be dependant in part on the distance of the original breeding pond from the new road.

However, the future design of tunnel, fence and substitute ponds will vary with site conditions and the criteria for success of a protective system may need to be refined following measurements of movement and genetic variation in populations once a variety of systems have been installed and compared. On a national basis, the co-ordination of system design would benefit from having a scientific overview and more applied studies would be valuable.

## RECOMMENDATIONS

### UK amphibian tunnel systems

Specific recommendations from the findings are i) existing tunnel and fence systems should be completed and maintained and studied by local education authorities or other interested parties using standardised techniques ii) sharing of experiences from different countries should be encouraged in order to speed up the development of system design iii) guidelines for the installation and monitoring of amphibian tunnels and fence systems should be drawn up, made widely available and refined each year as a form of reference at an early stage for those considering system construction.

### General

The importance of planning and designing adequate mitigation for wildlife during new road building or road upgrading include: reducing human death and injury and damage to vehicles, reducing death and injury of animals, reducing the genetic isolating and fragmentation effects of the road on animals (and plants), the retaining of corridors for animals with large home ranges and the avoidance of the loss or decline in size of animal populations. Successful projects can bring about appreciation by both the general public and professional road planners and builders. Poorly thought out or built systems however will appear wasteful and lower the sense of achievement in those involved with all aspects of scheme development, construction, maintenance and monitoring.

The concept of designing roads so that they minimise the visual impact on the landscape is not new, although the general concept may not always fit well with protecting all species. The position of a road in the landscape will cause differing effects according to the particular animals or plants influenced by its location. In terms of amphibians and reptiles, road positions low in valleys and close to rivers and wetlands on the valley basin may heighten the level of contact.

Small animals with a body weight under 5000g are less likely to cause vehicles to crash. Large domestic animals may be the primary cause of collision especially in Asia and Africa. In Africa, large heavy wild mammals such as antelope, giraffe, zebra, rhino, buffalo and elephant may walk or run across roads both during the day and at night. In Europe and the USA, elk, (moose) and deer cause most of the human fatalities in vehicle collisions. However, it was the death of a driver skidding upon seeing a mass migration of amphibians in Bavaria (Germany) that increased concern and contributed to pressure for measures to be taken to consider amphibian tunnel systems in Germany and Switzerland.

The reduction of death and injury of animals, reduction of genetic isolation and other fragmentation effects discussed earlier apply to many other species. Animals with much larger home ranges may need particular attention and it may be possible to categorise species in terms of the level of acceptable mitigation and this may vary greatly between species.

The following general recommendations are aimed at advancing studies of the problem and taking remedial action.

- With respect to road warning signs, involvement of volunteer 'rescues', road closures and similar activities, it should be remembered that the involvement of local people in the process is important. This often needs careful control and safety measures due to the proximity of fast moving traffic. Safe human access to places where animals migrate across roads is an important safety consideration.

- Vehicle speeds greatly influence the damage caused during animal-vehicle collision and the likelihood of a collision occurring. Serious consideration should be given to adjusting speed limits on roads in order to reduce probability of collision. With advances in technology, speed limits could even be varied according to time of day, as many collisions occur at dusk and dawn, or only at night. Such measures may also increase efficiency in petrol and diesel use in vehicles as car performance is often most efficient at 50-80 kph. Such policies would complement reduction of emissions of greenhouse gases and other pollutants that would occur through controlling (mostly lowering) speed limits.

- The production of Quality Standards guidelines for tunnel installation has occurred in some German Landers and Swiss Cantons for several years and is an area where the development of suitable texts is needed elsewhere. Proposals for promoting strategic approaches have been developed in the USA as a result of the State of Florida, Department of Transportation conferences that were held in 1996 and 1998. The review by Scott Jackson and Curtace Griffin 'Toward a Practical Strategy for Mitigating Highway Improvements on Wildlife' gives seven specific recommendations for strategic approaches to addressing wildlife mitigation issues. The potential for groups of countries/states/regions with similar species and habitats to promote a collective response to issues seems appropriate.

- Legislation has been a significant driving force for system construction and this is likely to continue. For this reason, better guidelines (see above), reinforced by more detailed legislative/semi-legislative frameworks could be advantageous. Such measures would involve national laws, regional directives and regulations, international conventions or other protocols and procedures.

- The broadest possible approach may include consideration of the need for new road construction (Luukkainen, 1989) in the light of global climate and carbon burning issues. It may be beneficial to define more precisely the level of environmental damage that is caused by new roads and road upgrading. New roads are often the most damaging to natural habitats and the development of sensitivity zones in each country might be an objective. It might even be possible to set quotas to limit the building of new or upgrading of older, roads in certain biogeographical zones. This might be especially worthwhile in the most sensitive countries/regions. An inter-government monitoring and reporting network may be necessary. Such a system might enable a better understanding of the level of existing and future threat to natural areas and wildlife from roads and offer a future information feedback and control mechanism relating the global exploitation of sensitive areas in all regions.

## ACKNOWLEDGEMENTS

Information on amphibian tunnel and fence systems was supplied on behalf of a range of organisations installing systems with funding provided by the Department of Transport, (Highways Agency, Department of Transport, Environment and the Regions). County Councils involved include Buckinghamshire, East Riding and North Yorkshire, West Sussex, Norfolk, Leicestershire, Cheshire and Suffolk. Chichester District Council, Barnsley Metropolitan Borough Council and Broxtone Borough Council also assisted with information. Details were also supplied by individuals and reports from the following organisations: ACO Cranfield Ltd, Fauna and Flora International, Persimmon Homes Ltd, Ringway Ltd, The National Trust, The Cambridgeshire Wildlife Trust, Tarmac Ltd, Wyeth Laboratories and Halcrow UK. Those providing information on amphibian tunnel systems included: A.S. Cooke, R. Davis, K. Hankins, M. Hannan, A. Harvey, W. Horsfall, M. Larder, D. Lloyd, I. Marshall, R. Oldham, W. Seale, A. Seddon, A. Simpson, P. Smith, D. Sneller, D. Whiteley, A. Wilson. Project information was compiled with the technical assistance of Herpetofauna Consultants International (C.L. Beckett). J.A. Burton and J.P. Foster provided comments on the draft text. J. Barr assisted with additional information.

Thanks are also due to the ACO Group who have supported the development of new technology in amphibian tunnel and fence systems: H.J. Ahlmann, R. Hill, P. Webb, D. Humphries and K. Hankins. This manuscript is dedicated to all those who have had a large herd of giraffe charge towards them at night while driving on the main road between Nairobi and Mombassa.

## REFERENCES

**Anderson P. 1993.** *Roads and nature conservation. Guidance on impacts, mitigation and enhancement.* Peterborough: English Nature.

**Anon. 1992.** Amphibian conservation: Manual for protecting amphibians on Roads Stuttgart: Ministry of Works. (In German). 59.

**Beebee TJC. 1996.** *Ecology and Conservation of Amphibians.* London: Chapman and Hall.

**Bernardino FS Jr, Dalrymple DJ. 1992.** Seasonal activity and road mortality of the snakes of the Pa-hay-okee wetlands of Everglades National Park, USA. *Biological Conservation.* **62:** 71-75.

**Berris L. 1997.** The importance of the ecoduct at Terlet for migrating mammals. 1997. In: Canters K. ed. *Habitat Fragmentation & Infrastructure, proceedings of the international conference on habitat fragmentation, infrastructure and the role of ecological engineering.* Delft: Ministry of Transport, Public Works and Water Management. 418-420.

**Boarman WI, Sazaki M. 1996.** Highway mortality in desert tortoises and small vertebrates: success of barrier fences and culverts. In: Evink GL, Garrett P, Zeigle D, Berry J. eds. *Trends in Addressing Transportation Related Wildlife Mortality,* Proceedings of the Transportation related wildlife mortality seminar. Tallahassee: State of Florida Department of Transportation. FL-ER-58-96.

**CODA. 1992.** *First Conference on the study and prevention of vertibrate mortality on roads.* Madrid: CODA. 3 vols. (in Spanish) 432.

**Devaux. 1992.** L'Autoroute A57: des engagements tenus. *La Tortue.* **19:** 5-7.

**Dexel R. 1989.** Investigation into the protection of migrant amphibians from the threats from road traffic in the Federal Republic of Germany: a summary. In: Langton TES. ed. *Amphibians and Roads, proceedings of the toad tunnel conference.* Shefford: ACO Polymer Products. 43-49.

**Fahrig L, Pedlar JH, Pope SE, Taylor PD, Wegner JF. 1995.** Effect of road traffic on amphibian density. *Biological Conservation.* **73:** 177-182.

**Foster ML, Humphrey SR. 1995.** Use of highway underpasses by Florida panthers and other wildlife. *Wildlife Society Bulletin,* **23 (1):** 95-100.

**Fowle SC. 1996.** Effects of roadkill mortality on the western painted turtle (*Chrysemys picta bellii*). In: Evink GL, Garrett P, Zeigler D, Berry J. eds. *Trends in Adddressing Transportation Related Wildlife Mortality.* Proceedings of the transportation related wildlife mortality seminar. Tallahassee: State of Florida Department of Transportation. FL-ER-58-96.

**Froglife. 1996.** *Toad patrols: a survey of voluntary effort involved in reducing road traffic - related amphibian mortality in amphibians.* Halesworth: Froglife Conservation Report No.1.

**Gittins SP, Parker AG, Slater FM. 1980.** Population characteristics of the common toad (*Bufo bufo*) visiting a breeding site in mid-Wales. *J. Anim Ecol.* **49:** 161-173.

**Gittins SP. 1983a.** Population dynamics of the common toad (*Bufo bufo*) at a lake in mid-Wales. *J. Anim Ecol.* **52:** 981-988.

**Gittins SP. 1983b.** Road casualties solve toad mysteries. *New Scientist.* February 24th: 530 - 531.

**Gittins SP. 1983c.** The breeding migration of the common toad (*Bufo bufo*) to a pond in mid-Wales. *Journal of Zoology, London.* **199:** 555-562.

**Grossenbacher K. 1981.** Amphibien und Verkehr. Koordinationsstelle Fur Amphibien - und Reptilenschutz in der Schweiz. *Publikation Nr 1. 2. Auflage. 25.*

**Hels T. 1999.** Effects of roads on amphibian populations. [PhD thesis] Denmark: National Environmental Research Institute.

**Hunt A, Dickens HJ, Whelan RJ. 1987.** Movement of mammals through tunnels under railway lines. *Aust. Zool.* **24(2):** 89-93.

**Jackson SD. 1997.** Proposed Design for an Amphibian and Reptile Tunnel. [Unpublished Manuscript].

**Jackson SD, Griffin CR. 1998.** Toward a Practical Strategy for Mitigating Highway impacts on Wildlife. *Proceedings of the International Conference on Wildlife Ecology and Transportation.* (ICOWET).

**Jackson SD. 1996.** Underpass Systems for Amphibians. In: Evink GL, Garrett P, Zeigler D, Berry J. eds. *Trends in Addressing Transportation Related Wildlife Mortality:* Proceedings of the transportation related wildlife mortality seminar. Tallahassee: State of Florida Department of Transportation. FL-ER-58-96.

**Jackson SD. & Tyning TF. 1989.** Effectiveness of drift fences and tunnels for moving spotted salamanders *Ambystoma maculatum* under roads. In: Langton TES. ed. *Amphibians and Roads.* Proceedings of the Toad Tunnel Conference. Shefford: ACO Polymer Products. 93-99.

**Jackson SD, Marchand MN. 1998.** Use of a Prototype Tunnel by Painted Turtles (*Chrysemys picta*) [Unpub. note.] Amherst: Department of Forestry and Wildlife Management, University of Massachusetts.

**Langton TES. 1989a.** Amphibians and Roads, proceedings from a study of a drift fence and tunnel system at Henley-on-Thames, Buckinghamshire, England. In: Langton TES. ed. *Amphibians and Roads.* Proceedings of the Toad Tunnel Conference. Shefford: ACO Polymer Products. 145-152.

**Langton TES. 1989b.** *Amphibians and Roads.* Proceedings of the toad tunnel conference. Shefford: ACO Polymer Products. 202.

**Luukkainen H. 1989.** Animal subways - view of an animal protectionist and green politician. In: Langton TES. ed. *Amphibians and Roads.* Proceedings of the Toad Tunnel Conference. Shefford: ACO Polymer Products.

**Mader HJ. 1984.** Animal habitat isolation by roads and agricultural fields. *Biological Conservation.* **29:** 81-96.

**Oxley DJ, Fenton MB. Carmody GR. 1974.** The effects of roads on population of small mammals. *Journal of Applied Ecology.* **11:** 51-59.

**Reh W. 1989.** Investigations into the influence of roads on the genetic structure of populations of the common frog *Rana temporaria.* In: Langton TES ed. *Amphibians and Roads.* Proceedings of the Toad Tunnel Conference. Shefford: ACO Polymer Products. 101-103.

**Reh W, Seitz A.** The influence of land use on the genetic structure of populations of the common frog *Rana temporaria. Biological Conservation.* **54:** 239-249.

**Schlupp I, Podloucky R. 1994.** Changes in breeding site fidelity: a combined study of conservation and behaviour in the common toad *Bufo bufo. Biological Conservation.* **69:** 285-291.

**Scottish Office. 1998.** *Cost Effective Landscape: Learning from Nature.* Landscape Design and Management Policy. A Roads, Bridges and Traffic in the Countryside Initiative. Edinburgh: Scottish Office.

**Stolz F-M, Podloucky R. 1983.** Krötentunnel als SchutzmaBnahme F• r wanderade Amphibien, dargestellt am Beispiel von Niedersachsen Informationsdienst Maturschurtz Nrl. Neidersächsisches *Landesverwaltungsamt - Fachebehörde fur Naturschutz.*

**Trombulak SC, Frissell CA. 2000.** Review of ecological effects of roads on terrestrial and aquatic communities. *Conservation Biology.* **14:** 18-30.

**van Gelder JJ. 1973.** A quantatative approach to the mortality resulting from traffic in a population of *Bufo bufo* L. *Oecologia (Berl.).* **13:** 93-95.

**van Leeuwen BH. 1982.** Protection of Migrating common toad (Bufo bufo) against Car Traffic in the Netherlands. *Environmental Conservation.* **34.**

**Way M. 1970.** Roads and the Conservation of Wildlife. *The Journal of the Institute of Highway Engineers.* July: 5-11.

**Yanes M, Velasco JM, Suarez F. 1995.** Permability of roads and railways to vertebrates: the importance of culverts. *Biological Conservation.* **71:** 217-222.

# 21

# Disturbance by traffic as a threat to breeding birds: evaluation of the effect and considerations in planning and managing road corridors.

## R REIJNEN, R FOPPEN, G VEENBAAS & H BUSSINK

Wildlife considerations in planning and managing road corridors recognise that habitat loss is probably the most important factor in reducing the population size of breeding birds. Although motorised traffic kills large numbers of birds annually, it is has been concluded repeatedly that road kills do not exert a significant pressure on the population size (e.g. Ellenberg *et al.*, 1981; Leedy & Adams, 1982; Bernard *et al.*, 1987; Bennett, 1991) However, little attention has been given to the effect of disturbance by traffic. Recently, evidence has shown that this effect can be very important in affecting breeding bird populations in many species of very different habitat types (Reijnen *et al.* 1995, 1996; Reijnen & Foppen, 1995).

The aim of this chapter is to stress the importance of considering the 'disturbance' effect in planning and managing road corridors. We discuss the following points.

- Evidence illustrating the nature and extent of the disturbance effect in breeding birds.
- Exploration of the dimensions of the problem.
- Possible measures to reduce the effects.
- General considerations for application.

## BIRD POPULATION RESPONSE TO DISTURBANCE BY TRAFFIC

### The effects on the breeding density

In earlier studies depressed densities of breeding birds adjacent to roads were found in only a few species that belong to quite different

**Table 1.**   Effects of road traffic on the density of breeding bird species of different taxonomic groups in The Netherlands.

| Taxonomic group | Number of investigated species in woodland | | Number of investigated species in agricultural grassland | |
|---|---|---|---|---|
| | Total | Affected | Total | Affected |
| Anatidae | 1 | - | 4 | 2 |
| Accipitridae | 1 | 1 | | |
| Phasianidae | 1 | 1 | | |
| Rallidae | | | 1 | 1 |
| Haematopodidae | | | 1 | 1 |
| Charadriidae | | | 1 | 1 |
| Scolopacidae | 1 | 1 | 2 | 1 |
| Columbidae | 3 | 2 | | |
| Cuculidae | 1 | 1 | | |
| Picidae | 3 | 2 | | |
| Alaudidae | | | 1 | 1 |
| Motacillidae | 1 | 1 | 2 | 1 |
| Troglodytidae | 1 | 1 | | |
| Prunellidae | 1 | - | | |
| Turdidae | 4 | 3 | | |
| Sylviidae | 8 | 8 | | |
| Muscicapidae | 2 | 1 | | |
| Aegithalidae | 1 | - | | |
| Paridae | 6 | 4 | | |
| Sittidae | 1 | - | | |
| Certhidae | 1 | 1 | | |
| Oriolidae | 1 | 1 | | |
| Corvidae | 3 | 2 | | |
| Sturnidae | 1 | - | | |
| Fringillidae | 2 | 2 | | |
| Emberizidae | 1 | 1 | | |

Nomenclature follows Voous (1973, 1977), Cited in Reijnen *et al.* (1995, 1996), Reijnen & Foppen (1995) taxonomic groups, such as warblers, waders and grouse (Clark & Karr, 1979; Ferris, 1979; van der Zande et al., 1980; Adams & Geis, 1981; Illner, 1992a). Since the total number of investigated species in these studies was rather small, it is difficult to interpret whether the density-depressing effect of roads is a common phenomenon, or not. However, from the results given in the recent papers of Reijnen et al. (1995, 1996) and Reijnen & Foppen (1995) it can be concluded that the effect is widely spread in breeding birds. Along heavily travelled roads in The Netherlands 33 òf the 45 investigated bird species in woodland and 7 of the 12 investigated species in agricultural grassland showed an effect. The affected species cover almost all taxonomic groups present in these data (Table 1).

Rough estimations of disturbance distances and density reductions over these distances in some of the earlier studies indicate that the effect might also be important quantitatively. For two wader species in open field habitat, van der Zande et al. (1980) estimated disturbance distances ranging from 625 m for a secondary road to 2000 m for a busy highway. Tetraonid species in woodland were disturbed up to a distance of 500 m near relatively quiet highways. In both studies the density reduction in the disturbed zone amounted to 50% or more.

Accurate quantitative data of many species from the recent studies of Reijnen et al. (1995, 1996) and Reijnen & Foppen (1995) show, that such large disturbance distances and high density reductions are not exceptional. They quantified the relationship between traffic load and density with regression by using a threshold model with traffic noise (in dB(A)) as the explanatory variable (see for explanation Reijnen et al., 1995; Figure 1). Disturbance distances were derived from these equations by transforming the threshold value in dB(A) into distance from the road (m). In estimating disturbance distances we disregard regressions in which a threshold is absent. The lowest noise level then represents the threshold and this would lead to very large and probably unrealistic disturbance distances. Following these assumptions the range of threshold values in dB(A) for species and for all species combined become very similar in both open grassland and woodland (Table 2). Estimated disturbance distances at 50,000 vehicles per day vary from 75 to 930 m for grassland breeding birds and from 60 to 810 m for woodland breeding birds. For the density of all species combined the estimated disturbance in open grassland is 560 m and in woodland 365 m (Table 3). At a traffic density of 10,000 vehicles per day the estimated disturbance distances in open grassland are four to five times lower and in woodland almost three times (Table 3).

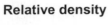

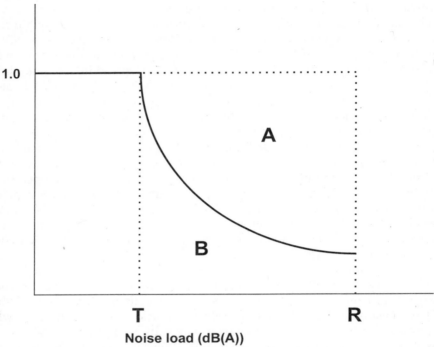

**Figure 1.** Threshold model for relative breeding density of birds against traffic noise, where T is the threshold value and R the value at the roadside. The decrease factor of the density = area of A/ (area of A+B).                                                                              Source: Reijnen *et al.*, 1995.

**Table 2.**   Threshold values in dB(A) for breeding birds. Below these values the density is not affected.

| Habitat | Species | All species combined |
|---|---|---|
| Woodland | 36-58 | 42-52 |
| Open grassland | 43-60 | 47 |

**Table 3.**    Maximum size of the zone adjacent to main roads (in m) that has a reduced density of breeding birds, when probably unrealistic values are not considered.

| Habitat | Species | | All species combined | |
|---------|---------|---------|---------|---------|
|         | 10,000 cars/day | 50,000 cars/day | 10,000 cars/day | 50,000 cars/day |
| Open grassland | 230 | 930 | 120 | 560 |
| woodland | 305 | 810 | 125 | 365 |

Based on results of Reijnen *et al.* (1995, 1996) and Reijnen & Foppen (1995)

The reduction of the density over the disturbance distances varies greatly between species, but is never less than 30%. In both habitat types, several species even show a density reduction of almost 100%. This means that dense traffic, in particular, can cause an important reduction in the number of species. Because many species are affected, there is also a significant reduction of the total density; in open agricultural grassland, 39% and in woodland, 35%.

In some previous studies higher densities of breeding birds close to roads were also found (Clark & Karr, 1979; Ferris, 1979; Adams & Geis, 1981). However, this can be explained by certain habitat conditions being much more favourable close to roads than farther away and therefore should not be interpreted as a positive effect of traffic (see Reijnen *et al.*, 1995). There are no indications that species might be favoured because of lack of competition (increase of density if there is an effect on related species) or better feeding conditions due to the presence of road victims (Reijnen *et al.*, 1995). This is in accordance with the fact that total bird density is also reduced.

## Possible causes and mechanisms

Reijnen *et al.* (1995) found almost no effects on species along roads when the noise load due to traffic was relatively low. This indicates that the presence of a road *per se* is not very important in affecting densities of breeding birds. Also, it is not probable that mortality due to collisions is an important causal factor. Although the numbers of road victims can be rather large (e.g. Adams & Geis, 1981; Hodson & Snow, 1965, Füllhaas *et al.*, 1989; van den Tempel, 1993), it was assumed that they are in general not sufficient to cause a significant increase of the total mortality of species (Leedy & Adams, 1982; Ellenberg *et al.*, 1981; Reijnen *et al.*, 1995). Support for this assumption is given by Reijnen & Foppen (1994), who observed equal survival rates of male willow warbler's *Phylloscopus trochilus* close to a busy highway and in areas at a distance of several hundred meters. Only for owls, in particular barn owl *Tyto alba*, road mortality might influence population size significantly (Joveniaux, 1985; Illner, 1992b; van den Tempel, 1993).

This implies that possible causes of decline will be related to emission of matter and energy by road traffic, such as pollution, visual stimuli and noise (van der Zande *et al.*, 1980; Leedy & Adams, 1982; Ellenberg *et al.*, 1981). Reijnen *et al.* (1995) show evidence that in woodland, noise is probably the most critical factor in causing reduced densities close to roads. In regression analysis using noise and visibility of vehicles as response variables, noise appeared to be the best, and in many species, the only predictor of observed depressed densities close to the road. A reduction of the total density could only be explained by noise. Furthermore, they made it plausible that other possible causes, such as pollution and visual stimuli, are not very important, because they operate at a very short distance from the road and have in all probability a limited effect. In open landscape, however, an effect of visual stimuli cannot be excluded for certain. Here, visual stimuli reach much farther than in wooded areas and breeding birds might respond differently (Reijnen *et al.*, 1996). On the other hand, a study of Illner (1992a) showed that in the absence of visual stimuli (a road bordered by hedgerows) grey partridge *Perdix perdix* densities in open arable farmland were still heavily depressed up to several hundred metres from busy highways. Also, breeding birds of open grassland and woodland respond very similarly to disturbance by traffic (Table 3). This indicates that noise is also the most critical factor in open landscapes.

Very little is known about how noise causes reduced densities of breeding birds. For the willow warbler it has been shown that close to a highway many males experience difficulties in attracting or keeping a female, and because of the lack of reproductive success move out of the road zone in the following year (Reijnen & Foppen, 1994; Foppen & Reijnen, 1994). This could point to distortion of the song of males as a possible mechanism (cf. Reijnen & Foppen, 1994). However, there is some evidence that disturbance of the vocal communication between birds is probably not a general mechanism in causing reduced densities (Reijnen *et al.*, 1995). An alternative, or more likely a supplementary explanation is that birds avoid areas close to roads because of stress (Reijnen *et al.*, 1995; Illner, 1992a).

Figure 2 summarises the probable relationship between traffic and density of breeding birds.

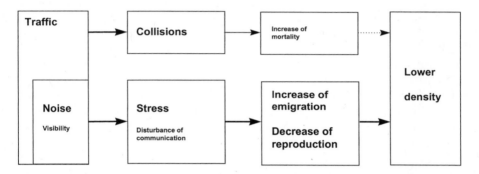

**Figure 2**. Probable relationship between road traffic and density of breeding birds.


## Effects on habitat quality in relation to density

There is much evidence that the reduction of the density is related to a reduction of the habitat quality (Reijnen & Foppen, 1994; Reijnen & Foppen, 1995). On the other hand it is known that density is not always a good indicator of habitat quality and might be even misleading (Fretwell, 1972; van Horne, 1983). In several territorial bird species it has been shown that, when overall density is high, less-preferred habitat is more strongly occupied than when overall density is low (Kluyver & Tinbergen, 1953; Glas, 1962; O'Connor & Fuller, 1985). Similar relationships were found between habitats close to roads and habitats further away (Reijnen & Foppen, 1995).

This means that the size of zones adjacent to roads, that have a lower quality due to disturbance by traffic, can easily be underestimated when it is based on density data (Figure 3). The study of Reijnen & Foppen (1995) indicates that in woodland birds such underestimation can be substantial, since many species only had a depressed density in years with a relatively low overall density. In consequence, species that did not show a change in density, still might be affected by traffic.

## Consequences for breeding population

Of the effects of disturbance by traffic on breeding bird populations, the reduction of the habitat quality is probably the single most important factor. There are many indications that the size and persistence of breeding populations mainly depends on areas with a high quality of habitat (Wiens & Rotenberry, 1981; Bernstein *et al.*, 1991). Consequently, the greatest effect on the overall population size can be expected when disturbance causes major losses of high-quality areas. Also further degradation of low quality habitats can have some effect, because they may contribute significantly to the overall population size (Howe *et al.*, 1991). This is supported by the study of Foppen & Reijnen (1994), who found that there is a quantitatively important breeding dispersal flow of male willow warblers from highway-induced low-quality habitat to nearby high-quality habitat.

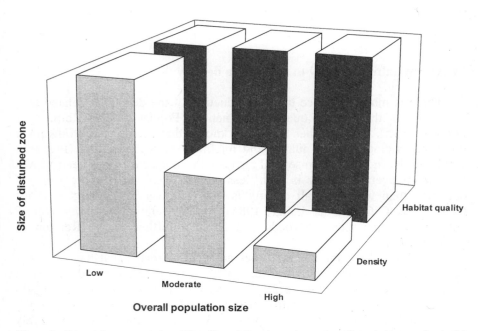

**Figure 3.** Schematic representation of the effect of disturbance by road traffic on habitat quality (solid) and density (hatched) of breeding birds in relation to overall population size.

Finally, one should also consider possible risks from accumulation effects. Breeding birds in The Netherlands, for example, suffer from many other environmental influences, of which eutrophication of ground water and surface water, ammonium deposition from agricultural emissions, lowering of the water table, and habitat fragmentation are considered to be of greatest importance (e.g. Canters & De Snoo, 1993; Bink *et al.*, 1994; Vos & Zonneveld, 1993) and this increases the risk of cumulative effects. The first three factors cause a decline of the habitat quality (or even a loss of habitat), which may affect the size of local populations. If the population size becomes very small there will be an increased risk of extinction due to chance demographic processes (Goodman, 1987; Shaffer, 1987), and habitat fragmentation can reinforce this process. It increases the risk of extinction due to chance demographic processes and it may reduce the chance that dispersal will rescue small local populations from extinction; eventually, this will affect whole network of populations (e.g. Opdam *et al.*, 1993; Verboom *et al.*, 1993). Evidence for these effects has been shown in several studies (see Opdam, 1991; Opdam *et al.*, 1993; 1995).

## EXPLORATION OF THE DIMENSIONS OF THE PROBLEM

### In The Netherlands

The Netherlands has a dense network of main roads with high traffic densities varying from circa. 8,000 to more than 140,000 vehicles per day (Anonymous, 1988, 1993b) (Figure 4). This network of main roads adsorbs the majority of the traffic volume on all paved roads outside the urbanised areas (Anonymous, 1992), and therefore it accounts for most of the effects of traffic on breeding bird populations.

The effect of disturbance by roads on breeding bird populations was quantified for woodland birds and for birds of moist and wet grasslands. Many of grasslands considered, in particular in the north and west of The Netherlands, are well known for their rich communities of so-called 'meadow birds', which are of international importance (Beintema, 1986). The most typical species is the black-tailed godwit *Limosa limosa*, of which 80-90% of the European population nests in these grasslands (van Dijk *et al.,* 1989).

Data on the distribution of moist and wet grasslands were taken from Bakker *et al.* (1989), which reflect the pattern of 1986. This data set also gives the distribution of the important sites for meadow birds, which are characterised by the presence of many species and high densities. For the pattern of woodlands we used a GIS database named the 'Bodemstatistiek' by the Dutch V Statistics Netherlands (CBS 1989).

We focused on the disturbance distances for all species combined, and for the most sensitive species. The analysis was carried out for the situation in 1993 and for a projected scenario in 2010 (Farjon et al. 1997), which is mainly based on an ongoing increase in the traffic density.

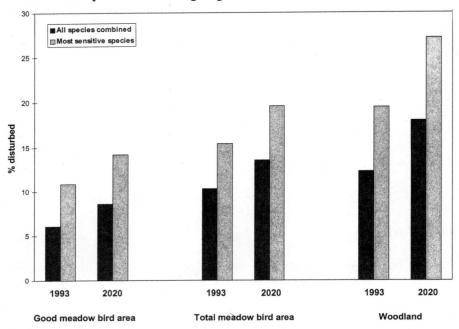

**Figure 4.** Relative size of disturbed zone for breeding birds by road traffic in The Netherlands in 1993 and for a scenario in 2020. Only roads with >8000 vehicles are considered.

Traffic data for 1993 were available in reports of the national and provincial authorities. Calculation of disturbance distances was carried out according to Reijnen & Foppen 1995, Reijnen *et al.* (1995) and Reijnen *et al.* (1996), and the size of the disturbed area was estimated by using overlay-techniques in a GIS system (ARCINFO).

Moist and wet grasslands suitable for meadow birds cover 885,110 ha, of which 146,441 ha can be considered as good meadow bird area. In 1993 the size of the disturbed zone by road traffic was considerable. For all species combined 86,120 ha (10.3%) of the total grassland area was

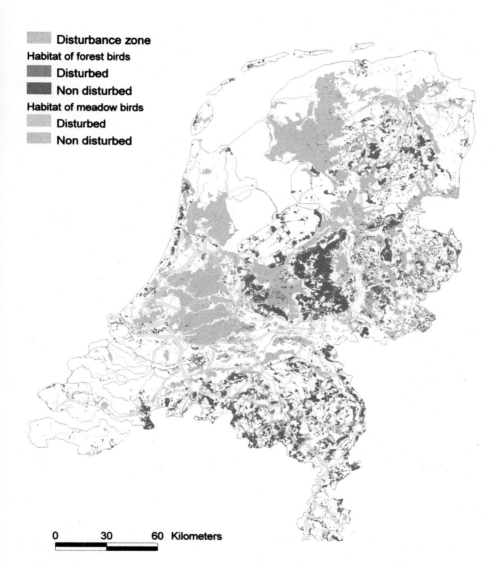

**Figure 5.** Pattern of disturbed area for breeding birds by road traffic in The Netherlands in 1993. Only roads with >8000 vehicles are considered.

disturbed and 7,040 ha (6.1%) of the area which was most suitable for meadow birds was disturbed. For the most sensitive species, the black-tailed godwit, the disturbed zones covered 126,145 ha (15.4%) and 9,010 ha (10.8%) respectively. (See Figure 4 and Figure 5)

Woodland covers 331,730 ha in total. In 1993 40,310 ha (12.2%) was disturbed for all species combined and 64,750 ha (19.5%) for the most sensitive species, the cuckoo *Cuculus canorus*. (See Figure 4 and Figure 5)

In both habitat types observed reductions in the density of birds in the disturbed zone are very significant. For all species combined these reductions are 39% in woodland and 35% in moist and wet grassland, and for the most sensitive species >90% and 47% respectively. In combination with the large area of disturbed habitat this points to a serious loss of population in many species.

Because of the ongoing strong increase of traffic densities, the effect will only become more important. By 2020 the disturbed zones along roads might be 30-50% larger than in 1993 (Figure 4), so, it is very important to take measures which avoid and reduce the effects.

**In the United States**

Based on the studies in The Netherlands, Forman (2000), made an estimate of the total ecological effect of the U.S. system of public roads. All road lengths in rural areas and 25% of road lengths in urban areas are roughly estimated to be near natural ecosystems. Natural ecosystems include agricultural land but exclude built areas. Calculations indicate that 19% of the total area of the United States is directly affected ecologically by roads and associated vehicular traffic. Primary and secondary roads have the same total ecological effect. Excluding Alaska (which has very few roads in a large area) and Hawaii means that 22% of the contiguous United States is estimated to be ecologically altered by the road network.

**POSSIBLE MEASURES TO REDUCE THE EFFECTS**

Because noise is probably the most critical factor, one can expect that a reduction of the noise load will reduce the effects on the density of birds. Although further experimental evidence is needed, there is no reason for not applying this knowledge now. Measures that reduce the noise load, such as the construction of walls of earth or concrete materials, will also reduce or eliminate the other traffic related factors.

A reduced density of bird populations adjacent to roads can also be compensated for by developing new favourable habitats outside the disturbed zone. Improvement of the habitat quality within this zone, in general, will have a limited effect, since the observed density reductions were very large.

However, since the reduction of the density decreases with increasing distance from the road, it might have some benefits in the part of the disturbed zone far from the road.

## Reducing effects along existing roads

A sufficient reduction of the noise load along roads in order to reduce the effects of traffic on breeding bird populations can only be achieved by constructing noise barriers. However, to obtain a substantial reduction, these barriers need to be very long and high. Moreover, in open areas, such barriers may also act as a source of disturbance themselves, because many birds of open grassland avoid the vicinity of hedgerows, wooded banks and dikes for up to several hundred meters (e.g. Klomp, 1954; van der Zande et al., 1980; Altenburg & Wymenga, 1991). So, this measure seems only appropriate for major problems (large disturbance distances in important areas for breeding birds). However, in constructing noise barriers for birds, another factor one should take into account is that they can hamper movements of other animal species. This might present fewer problems if under- or overpasses for wildlife are present. Openings in noise barriers at the ground level are probably not very effective, since, for small animals in particular, the barrier effect of the road itself seems much more important (e.g. Oxley, 1974; Mader, 1984; Mader et al., 1990). However, they might be useful for animals to escape from the road area.

Compensation for population loss along roads by creating new habitats outside the disturbed zone, has the disadvantage that it will usually take many years before the habitat is fully developed. Moreover, in The Netherlands with a dense human population, this application is probably limited. At present, re-allotment plans afford the best opportunity, in particular with respect to birds of agricultural grasslands. Extension of agricultural management in areas outside the disturbed zone (which improves the habitat quality), can be compensated for by intensifying the agricultural management close to roads (cf. Reijnen et al., 1996). One should take care that such measures do not cause loss of other natural habitats of value in the disturbed zone, such as interesting plant communities.

As discussed before, improvement of habitat quality of disturbed areas near the road will not be of great value to bird populations since the reduction of the density of many species is very marked. Moreover, it can increase the number of road kills (van den Tempel, 1993). In view of this, one should also have low expectations of positive effects of roadside management for breeding birds in general. This does not mean that roadside management to maintain or improve natural habitats, is not effective from a conservation point of view. In particular, when roadside habitat comprises remnants of natural vegetation, it can be valuable for many plant and animal species, such as small mammals, butterflies, carabid beetles and other invertebrates (e.g. Bennett, 1991; Sykora *et al.*, 1993; Hochstenbach, 1993). If verges connect other habitat patches, they also can function as a corridor, which might have positive effects on the size and persistence of the overall population (see e.g. studies on carabids, Vermeulen, 1993, 1994). In large agricultural areas, the occurrence of many species, including birds, may depend on roadside habitat (see e.g. Bennett, 1991).

## Minimising effects when planning roads

The best way to minimise the effects of new roads is to avoid disturbance of important areas for breeding birds. This can be achieved by using knowledge of disturbance distances in the first phase of the EIA-procedure, to which all plans for main roads are subjected (at least in the Netherlands). In exploring possible solutions for transportation routes, one should take into account a sufficient distance from these areas, based on expected traffic densities. In general, 1,000 m to both sides of the road seems an adequate distance.

When effects are inevitable, one should consider mitigation measures. As discussed above, the development of natural areas in or adjacent to roadsides are generally not a useful strategy with respect to breeding birds.

## GENERAL CONSIDERATIONS FOR APPLICATION

In applying the recent knowledge on the relationship between traffic and breeding bird populations in spatial planning procedures we distinguish between, assessment of the size of the problem and selection of effective measures that reduce the effects.

The use of noise to quantify the traffic load has the advantage of describing the relationship between traffic and density in a rather universal way. Noise takes into account many characteristics of traffic that might be important in affecting breeding birds (such as number and size of vehicles, speed) and, as mentioned above, is probably the most critical factor. Furthermore, there are appropriate mathematical models to calculate the noise load along roads (Moerkerken & Middendorp, 1981; Huisman, 1990; see also Reijnen *et al.*, 1995) and basic data are readily available or can be easily measured.

The available data give adequate information to establish effects of traffic on main roads in woodland (deciduous and coniferous) and open agricultural grassland. Sampling plots were distributed all over The Netherlands and the range of traffic densities involved was rather large (3,000-75,000 vehicles per day). Furthermore, application of the results in other habitat types seems allowable. The species studied are representative for the whole group of breeding bird species in The Netherlands and many of them can also be found breeding outside woodland and open agricultural grassland. It is likely that the results are also applicable in areas outside The Netherlands that have a similar bird fauna.

However, in hilly areas the established relationships between noise and density might not be valid. In that case and in all other very different situations further research is needed.

Effect curves in which a threshold was absent resulted in very large and probable unrealistic disturbance distances and it is better to not use them. On the other hand, there is evidence the available data underestimates the size of the disturbed zone considerably. To reduce this risk, for the remaining effect curves, one can concentrate on the largest effect found for the total density and the effect on the most sensitive species. These effect curves also have relatively narrow confidence limits. A side-advantage of this approach is that it makes extrapolation to other habitat types more easy.

To favour the application of the present knowledge on the effects of traffic on breeding bird populations in spatial planning procedures related to main roads, such as EIA and in road management practice, we compiled a manual that makes application easier (Reijnen *et al.*, 1995).

# References

**Adams LW, Geis AD. 1981.** *Effects of highways on wildlife.* Report No. FHWA/RD-81/067 Washington: Office of Research, Federal Highway Administration, U.S. Department of Administration.

**Altenburg W, Wymenga E. 1991.** Beheersovereenkomsten in veenweiden; mogelijke effecten op vegetatie en weidevogels. *Landschap* **8**: 33-45.

**Anon. 1992.** *Statistiek van de wegen.* Voorburg/Heerlen: Centraal Bureau voor de Statistiek (CBS).

**Anon. 1993a.** *Verhardingsgegevens hoofwegennet ZOAB en cement-betonverhardingen; Jaarrapport 1993.* Delft: Dienst Weg- en Waterbouwkunde.

**Anon. 1993b.** *Verkeersgegevens jaarrapport 1993.* Rotterdam: Adviesdienst Verkeer en Vervoer, Ministerie van Verkeer en Waterstaat.

**Bakker JJ, van Dessel B, van Zadelhoff FJ. 1989.** *Natuurwaardenkaart 1988.* SDU uitgeverij, 's-Gravenhage.

**Beintema AJ. 1986.** Man-made polder in The Netherlands: a traditional habitat for shorebirds. *Colonial Waterbirds.* **9:** 196-202.

**Beintema AJ. 1991.** Breeding ecology of meadow birds (Charadriiformes); implications for conservation and management. [Ph.D. Thesis] State University of Groningen, Groningen.

**Bennett AF. 1991.** Roads, roadsides and wildlife conservation: a review. In: Saunders DA, Hobbs RJ, eds. *Nature conservation 2: the role of corridors* Chipping Norton: Surrey Beatty and Sons.

**Bernard JM, Lansiart M, Kempf C, Tille M. 1987.** *Routes et Faune Sauvage; actes du colloque.* SETRA, Colmar.

**Bernstein, C, Krebs JR, Kacelnik A. 1991.** Distribution of birds amongst habitats: theory and relevance to conservation. In: Perrins CM, Lebreton JD, Hirons GJM. eds. *Bird Population Studies, relevance to conservation and management.* Oxford: Oxford University Press.

**Bink RJ, Bal D, van den Berk VM, Draaijer LJ. 1994.** *Toestand van de natuur 2.* Wageningen: IKC-NBLF.

**Canters KJ, De Snoo GR. 1993.** Chemical threat to birds and mammals in The Netherlands. *Reviews of Environmental Contamination and Toxicology* **130:** 1-29.

**Clark WD, Karr JR. 1979.** Effects of highways on red-winged blackbird and horned lark populations. *Wilson Bulletin.* **91:** 143-145.

**Ellenberg H, Müller K, Stottele T. 1981.** Strassen-Ökologie. In: Walper KH, Ellenberg H, Müller K, Stottele T. eds. *Ökologie und Strasse.* Bonn: Broschürenreihe de Deutschen Strassenliga,.

**Farjon JMJ, Hazendonk NFC, Hoeffnagel WJC. eds. 1997.** Verkenning Natuur en Verstedelijking. Wageningen: Achtergronddocument 10 Natuurverkenning '97, Informatie- en kenniscentrum Natuurbeheer.

**Ferris CR. 1979.** Effects of Interstate 95 on breeding birds in northern Maine. *Journal of Wildlife Management.* **43:** 421-427.

**Fretwell SD. 1972.** *Populations in a seasonal environment.* Princetown: Princetown University Press.

**Foppen R, Reijnen R. 1994.** The effects of car traffic on breeding bird populations in woodland. II. Breeding dispersal of male willow warblers (*Phylloscopus trochilus*) in relation to the proximity of a highway. *Journal of Applied Ecology.* **31:** 95-101.

**Forman RTT. 2000.** Estimates of the area affected ecologically by the road system in the United States. *Conservation Biology.* **14:** 31-35.

**Füllhaas U, Klemp C, Kordes A, Ottersberg H, Pirmann M, Thiessen A, Tschoetschel C, Zucchi H. 1989.** Untersuchungen zum Strassentod von Vögeln, Säugetieren, Amphibien und Reptilien. *Beiträge Naturkunde Niedersachens* **42:** 129-147.

**Glas P. 1962.** Factors governing density in the chaffinch (*Fringilla coelebs*) in different types of wood. *Archives Neerlandaises de Zoologie* **13:** 466-472.

**Goodman D. 1987.** The demography of chance extinction. In: Soulé ME. ed. *Viable populations for conservation.* Cambridge: Cambridge University Press.

**Hochstenbach SMH. 1993.** *Keuze bij inrichting en beheer van wegbermen.* Delft: Vogelbescherming Nederland, Zeist and Rijkswaterstaat Dienst Weg- en Waterbouwkunde.

**Hodson NL, Snow DW. 1965.** The road deaths enquiry, 1960-61. *Bird Study* **12:** 90-99.

**Howe RW, Davies GJ, Mosca V. 1991.** The demographic significance of 'sink' populations. *Biological Conservation.* **57:** 239-255.

**Huisman W. 1990.** Sound propagation over vegetation-covered ground. [Ph.D. Thesis]. Nijmegen: Catholic University of Nijmegen.

**Illner H. 1992a.** Effect of roads with heavy traffic on grey partridge density (*Perdix perdix*) density. *Gibier Faune Sauvage* **9:** 467-480.

**Illner H. 1992b.** Road deaths of Westphalian owls: methodological problems, influence of road type and possible effects on population levels. In: Galbraith CA, Taylor IR, Percival S. eds. *The ecology and conservation of European owls.* Peterborough: Joint Nature Conservation Committee.

**Joveniaux A. 1985.** Influence de la misé en service d'une autoroute sur la faune sauvage. In: Bernard J-M, Lansiart M, Kempf C, Tille M. eds. *Actes du colloque Routes et Faune Sauvage.* SETRA, Colmar.

**Klomp H. 1954.** De terreinkeus van de kievit (*Vanellus vanellus*). *Ardea* **42:** 1-139.

**Kluyver HN, Tinbergen L. 1953.** Territory and the regulation of density in titmice. *Archives Neerlandaises de Zoologie* **10:** 265-289.

**Leedy DL, Adams LW. 1982.** *Wildlife considerations in planning and managing highway corridors.* Report No. FWHA-TS-82-212, Washington: Office of Research, Federal Highway Administration, U.S. Department of Administration.

**Mader HJ. 1984.** Animal habitat isolation by roads and agricultural fields. *Biological Conservation.* **29:** 81-96.

**Mader HJ, Schell C, Kornacker P. 1990.** Linear barriers to arthropod movements in the landscape. *Biological Conservation.* **54:** 209-222.

**Moerkerken A, Middendorp AGM. 1981.** *Berekening van wegverkeersgeluid.* Staatsuitgeverij, 's-Gravenhage.

**O'Connor RJ, Fuller RJ. 1985.** Bird population responses to habitat. In: Taylor K, Fuller RJ, Lack PC. eds. *Bird census and atlas studies: Proceedings of the VII international conference on bird census work and atlas work.* Tring: British Trust for Ornithology.

**Opdam P. 1991.** Metapopulation theory and habitat fragmentation: a review of holarctic breeding bird studies. *Landscape Ecology* **5:** 93-106.

**Opdam P, van Apeldoorn R, Schotman A, Kalkhoven J. 1993.** Population responses to landscape fragmentation. In: Vos CC, Opdam P. eds. *Landscape ecology of a stressed environment..* London: Chapman and Hall.

**Opdam P, Foppen R, Reijnen R, Schotman A. 1995.** The landscape ecological approach in bird conservation: integrating the metapopulation concept into spatial planning. *Ibis* **137:** 139-146.

**Oxley DJ, Fenton MB, Carmony GR. 1974.** The effects of roads on populations of small mammals. *Journal of Applied Ecology.* **11:** 51-59.

**Reijnen MJSM, Veenbaas G, Foppen RPB. 1995.** *Predicting the effects of motorway traffic on breeding bird populations.* Delft, Wageningen: Road and Hydraulic Engineering Division of the Ministry of Transport, Public Works and Water management/DLO-Institute for Forestry and Nature Research.

**Reijnen R, Foppen R. 1994.** The effects of car traffic on breeding bird populations in woodland. I. Evidence of reduced habitat quality for willow warblers (*Phylloscopus trochilus*) breeding close to a highway. *Journal of Applied Ecology.* **31:** 85-94.

**Reijnen R, Foppen R. 1995.** The effects of car traffic on breeding bird populations in woodland. IV. Influence of population size on the reduction of density close to a highway. *Journal of Applied Ecology.* **32:** 481-491.

**Reijnen R, Foppen R, ter Braak C, Thissen J. 1995.** The effects of car traffic on breeding bird populations in woodland. III. Reduction of density in relation to the proximity of main roads. *Journal of Applied Ecology.* **32:** 187-202.

**Reijnen R, Foppen R, Meeuwsen H. 1996.** The effects of traffic on the density of breeding birds in Dutch agricultural grasslands. *Biological Conservation.* **75:** 255-260.

**Räty M. 1979.** Effect of highway traffic on tetraonid densities. *Ornis Fennica* **56:** 169-170.

**Shaffer M. 1987.** Minimum viable populations: coping with uncertainty. In: Soulé ME. ed. *Viable populations for conservation.*

**Sykora KV, de Nijs LJ, Pelsma TAHM. 1993.** *Plantengemeenschappen van Nederlandse wegbermen.* Utrecht: KNNV.

**van den Tempel R. 1993.** *Vogelslachtoffers door het wegverkeer.* Zeist: Vogelbescherming Nederland.

**van der Zande AN, ter Keurs WJ, van der Weijden WJ. 1980.** The impact of roads on the densities of four bird species in an open field habitat - evidence of a long-distance effect. *Biological Conservation.* **18:** 299-321.

van Dijk AJ, van Dijk G, Piersma T, SOVON. 1989. Weidevogelpopulaties in Nederland. *Het Vogeljaar* 37: 60-68.

van Horne B. 1983. Density as a misleading indicator of habitat quality. *Journal of Wildlife Management.* 47: 893-901.

Verboom J, Metz JAJ, Meelis E. 1993. Metapopulation models for impact assessment of fragmentation. In: Vos CC, Opdam P. eds. *Landscape ecology of a stressed environment.* London: Chapman and Hall.

Vermeulen HJW. 1993. The composition of the carabid fauna on poor sandy roadside verges in relation to comparable open areas. *Biodiversity and Conservation.* 2: 331-350.

Vermeulen R. 1994. Corridor function of a road verge for dispersal of stenotopic heathland ground beetles *Carabidae. Biological Conservation.* 69: 339-349

Vos CC, Zonneveld JIS. 1993. Patterns and processes in a landscape under stress: the study area. In: Vos CC, Opdam P. eds. *Landscape ecology of a stressed environment.* London: Chapman and Hall.

Wiens JA, Rotenberry JT. 1981. Censusing and the evaluation of habitat occupancy. *Studies in Avian Biology* 6: 522-532.

The main part of this chapter is taken from:
**Reijnen R, Foppen T, Veenbaas G. 1997.** Disturbance by traffic of breeding birds: evaluation of the effect and considerations in planning and managing road corridors. *Biodiversity and Conservation.* 6 (4): 567-581.

Mainly reprinted pages: 567-573, 577-581. Reprinted figures 1, 2 & 3.
Original copyright notice: 0960-3115 © Chapman and Hall.
With kind permission from Kluwer Academic Publishers.

# 22

# Wildlife mortality: head-on collisions

J BARTON

The following paper was given over three years ago at a Linnean Society Symposium in March 1998. The information for both this paper and the original regional and national reports (Appendix I) are based on, was gathered by many hard pressed Wildlife Trust staff and volunteers. Some funding and help in kind was provided by World Wildlife Fund, Countryside Commission, English Nature, Countryside Council for Wales, Scottish Natural Heritage and the Royal Society for the Protection of Birds. However the bulk of the original data collection was essentially unfunded and reliant on the goodwill of people who had many other environmental issues to deal with. This means there are some gaps in coverage and some inconsistencies in the data. An example is where some Wildlife Trusts have measured the length of road that will impact on a site but others have not done this. To avoid confusion I have decided to limit most of the data shown in the Appendices at the end of this chapter to the number of sites affected.

Despite these concerns, the 1994-97 series of *Head on Collision Reports* still represent the most comprehensive study of the probable impacts of the then Roads Programmes on biodiversity and landscapes. They show the sheer magnitude of the problem at that time.

Since 1997-98 there has been no follow up work to assess the likely impact of subsequent Roads Reviews, mainly due to lack of funding and the increase in my own organisation's size and demands on my time.

Equally, the Government's change in attitude to road building slowed the whole process down. The restructuring of Government Departments and the Highways Agency, plus an increase in the strength of environmental protection as part of the decision-making processes have led to a distinct improvement in the situation. The New Approach to Trunk Road Appraisal (NATA) and Guidance on Methodology for Multi-Modal Studies (GOMMMS DETR 2000) are now accepted, as is the system of Multi-Modal Studies introduced to include all forms of transport as part of the solution to a transport problem. I sit on ORBIT – the Multi-Modal Study for the M25 – and can testify to the efforts to ensure all alternatives to roads are being considered.

However, the protected landscapes and biodiversity contained within them and other sites are still there. Many areas of habitat value are not even designated due to the lack of resources for surveys or the comparative abundance of that habitat, for example an area of rush pasture in Dumfriesshire may not be of significant value there, but would be hugely valuable in the context of Surrey. All of these areas within the supporting matrix of farmland make up our countryside.

Although the Roads Programme has been reorganised, many of the roads noted in our *Head on Collision Reports* are still active and some have been built. There will be huge impacts on our biodiversity if we continue to try to build our way out of traffic congestions. The 10 Year Plan and other Government initiatives may have lessened the threat but they have not removed it.

Consequently, I stand by the message in this chapter about the risks to our environment of road building. I would also ask that we seriously consider revisiting the database with the latest Roads Programme to be sure of the up-to-date picture. I hope that all the improvements in policy will have improved it but fear that may well not be so.


## INTRODUCTION

Since the Department of Transport published its controversial White Paper *Roads for Prosperity* in 1989, the debate over road building policy and its real environmental cost has continued to grow. More recently the Government has published a review of its road building programme in *Trunk Roads in England Review* (Department of Transport, 1994).

This chapter aims to contribute to the debate by providing a comprehensive assessment of the impact of road schemes on wildlife sites and protected landscapes in Britain.

In 1990, the Wildlife Trusts in South East England launched a Transport Campaign with the aim of highlighting the impact of road building on important wildlife sites. The resultant publication of *Head on Collisions, Road Building and Wildlife in South East England (1990)* gave the most comprehensive review ever made of the threats to wildlife sites from road building schemes in nine counties of South East England. This set the model followed in later reports throughout Scotland, Wales and the English regions between 1994-97. This chapter builds upon and updates the findings of these previous *Head on Collisions* to give a new assessment of the road building programme on important wildlife sites. It also gives a detailed appraisal of road schemes that threaten the protected landscape areas of Britain.

The information gathered for the production of this report is derived from the 46 Wildlife Trusts across Britain. Each Trust was asked to compile a schedule of proposed road schemes that impacted upon important wildlife sites and protected landscapes in the county or counties within their responsibility. To ensure consistency of approach, clear criteria were required to identify proposed road schemes, important wildlife sites and protected landscapes. The following definitions were adopted:

## PROPOSED ROAD SCHEMES

These were highlighted from the appropriate County Transport Policies and Programmes (TPP). This lists road schemes proposed within the county for the following five-year period. Road developments being promoted by both the Department of Transport (DTp) and the County Council are identified in each TPP. In several of the counties investigated additional schemes that were not included within TPP were also identified from Local and Structure Plans and direct consultations between Local Authorities and the Wildlife Trusts.

During the compilation of this information the Government published their review of the roads programme in *What Role for Trunk Roads in England* in 1997. Despite being heralded as a significant reduction in the road building programme the Review appears to have rearranged the existing road schemes rather than eliminating them altogether. For example in the South East, it withdrew only three of the previously planned road schemes and brought forward one (the upgrading of the M27).

## IMPORTANT WILDLIFE SITES

Sites considered to be of importance to wildlife have been identified by reference to widely accepted criteria for the assessment of such sites.

Sites of European significance to the conservation of birds are designated as Special Protection Areas (SPA). These are designated by the Department of the Environment in accordance with the 1979 EC Directive on the Conservation of Wild Birds. Internationally important wetlands are designated as Ramsar Sites by the Department of the Environment, in accordance with the Ramsar Convention on the Conservation of Wetlands of International Importance (1972). In many instances internationally important wetlands are also of importance to the conservation of wild birds and are consequently designated as both Ramsar and SPA sites.

SSSI form a nation-wide network of the most valuable habitats in Britain. In England, they are identified by the Government's statutory advisors on nature conservation, English Nature, and are notified in accordance with the Wildlife and Countryside Act 1981. This role is filled by the Countryside Council for Wales and Scottish Natural Heritage in those countries. All SPA and Ramsar sites are also notified as SSSI.

Within each county, County Wildlife Sites (CWS), Sites of Nature Conservation Importance (SNCI) or Countryside Heritage Sites (CHS) have been identified by the Local Planning Authority in partnership with the Wildlife Trust and other nature conservation bodies. These are sites which are considered important for the conservation of species and habitats within each county, supporting and extending the value of the SSSI network. Many counties have adopted specific policies for the protection of such sites through Local and Structure Plans. Due to the non-statutory and county-based designation of CWS, there is no guarantee of consistency in the quality of these sites between counties. However they are generally regarded as being of at least county importance to nature conservation.

Not all important wildlife sites have a formal designation and these are assigned to the category in the appendices at the back of the chapter. However, each wildlife site identified by Wildlife Trusts has been assigned to one of a number of habitat types. The importance of these wildlife habitats is described below.

**Salt marsh and mudflat**

Estuaries with tidal mudflats and fringing salt marshes are found all around Britain's coastline. Britain's coastal wetlands are often internationally important for wildlife, providing winter refuges and passage refuelling sites for many thousands of migratory wading birds and wildfowl. Many estuaries qualify for both Ramsar and SPA status.

**Ancient woodlands**

Ancient woods are those sites which have been wooded since at least 1600 AD, most are much older. They support a rich diversity of wild plants and animals dependent upon the stable environmental conditions provided by the long established woodland cover. Both the Government and their advisors on forestry, the Forestry Authority, have recognised the special environmental qualities of this habitat. Many ancient woods retain the natural mix of trees, shrubs and other species that originally colonised them.

Although the relative abundance of species and structure of the wood will have been modified by traditional woodland management, they represent the closest links we have with the original primeval and natural wildwood that once cloaked much of Britain.

Such sites are termed Ancient Semi-Natural Woodland (ASNW). Other ancient woods have been replanted, often with non-native coniferous species. The more natural mix of plants and animals found in an ASNW is often masked by the presence of these planted trees, and the nature conservation of such woods is damaged by such planting.

### Recent semi-natural woodland and scrub

Much of the woodland in Britain has evolved naturally over other habitats, commonly on the chalk downlands and lowland heaths of Southern England, or moorlands and other upland habitats of the North and West. Areas of recently developed woodland and scrub are often of considerable importance to wildlife even if they do not support some of the more specialist species found in ancient woodland or the other, longer established habitats they have replaced. The value to wildlife of scrub and recent woodland, in a mosaic with other habitats such as unimproved grassland, chalk downland and heathland, has only recently been fully appreciated.

### Rivers, streams, ditches and ponds

Areas of open water support a rich and distinctive range of wildlife. However, the rivers and streams of Britain have often been damaged by pollution and low flows caused by over abstraction. Many farm ponds and ditches have been infilled to improve agricultural efficiency. Those areas of open water that remain are threatened by ever-increasing disturbance from recreational use, roads and other development.

### Wet marshy grassland

Water meadows and grazing marshes were once a common feature of the river valleys and flood plains of lowland Britain. Many of these wetlands are internationally important for breeding wading birds and wintering waterfowl and qualify for SPA and Ramsar designation. Although many have been damaged by the application of fertiliser and herbicide, still more have been destroyed by land drainage and conversion to arable land. Only a few fragments of this once common wildlife habitat now survive intact.

## Old meadow grasslands

These are areas of long established wild flower rich meadow grassland that have been managed without the application of artificial fertiliser or herbicide. Grasslands of this type support an abundance of wild plants and animals. Britain has lost an estimated 95% of its wildlife rich hay and grazing meadow grasslands since the 1940s.

## Chalk downland

The downland of Southern and Eastern England supports an assemblage of wild plants and animals that are virtually confined to this habitat. Britain has lost an estimated 80% of its traditionally managed chalk grassland since the 1940s, mainly as a result of agricultural improvement.

## Heathland

Lowland heathland is an internationally rare habitat confined to the western fringe of Europe. It supports a characteristic assemblage of plants and animals, many of which are restricted to this habitat and thus, are also internationally rare and endangered. This includes birds such as the Nightjar, Dartford Warbler and Woodlark.

Britain contains some of the largest lowland heathland areas in the world, much of which qualifies for SPA designation. Unfortunately, an estimated 40% of Britain's lowland heathland has been destroyed since the beginning of this century by afforestation and agricultural improvement.

## Valley mire and bog

Valley mires and bogs are often found within heathland. Like heathland, they are internationally rare habitats supporting many uncommon and threatened plants and animals. Britain's largest system of valley mires and bogs is found in the New Forest, which is designated as a Ramsar site.

## Protected landscapes

As well as wildlife sites this study also looks at the impact of proposed road schemes on protected landscapes. Areas of Outstanding Natural Beauty

(AONB) are designated by the Countryside Commission. They represent the most important landscapes in England and Wales and in terms of the landscapes they protect, are considered to enjoy equal status with National Parks. Their protection is therefore no less important than that of the nature conservation sites.

## Identification of threatened sites

All sites threatened by road schemes were initially identified by the County Wildlife Trusts. These were critically examined before inclusion in this report. To be included, sites had to meet one of the following criteria:

- There was clear evidence that a proposed new road route would pass through an important wildlife site.

- Important wildlife sites were located immediately adjacent to existing road routes where there are proposals to widen or otherwise upgrade such routes.

In some instances more than one route has been proposed for a new road, these are all listed in the County Schedules. The summary statistics produced for each county and habitat draw on the most damaging of the route options, except where a preferred route has been identified.

A note of caution must be drawn here. In some counties some SSSI's are also CWS's. We have tried to eliminate double counting by careful scrutiny of the individual *Head on Collision Reports*. Where we could not be sure of SSSI status we have assigned the site to the Wildlife Site category. That is, we have underestimated the value of some sites that will be affected to avoid inflating the total numbers of sites.

## ASSESSMENT OF IMPACT

### Direct impacts

The construction of a new road through an important wildlife habitat or protected landscape has clear and dramatic impacts upon the habitat that is grubbed up or that is scarred. These direct impacts are relatively easy to quantify once the detailed design of the road, including its width, cutting locations and embankment sizes are known. These engineering design features can significantly increase or decrease the direct impact of a road scheme.

Unfortunately in most instances the design of a new road is not made public until well into the road planning process. As a consequence, it has only been possible to estimate the magnitude of direct environmental impact of many of the road schemes by giving the length of the habitat or landscape that is crossed by the road and the width of the proposed road, where this has been published. As not all the regional reports gave lengths we have not used this in this national summary paper.

In some instances the direct impact of widening and upgrading existing roads can be less than that caused by the construction of new roads. This is especially the case where proposals are to confine the widening to the existing carriageway width, but there are still impacts associated with this.

**Indirect impacts**

The construction of new roads and the enlargement of existing ones have a number of often very serious indirect impacts on habitats and species. These include habitat fragmentation, disruption of water supplies to and from wetlands, pollution of air and watercourses, disturbance of ground nesting birds and the death of badgers, barn owls, deer and other animals as road casualties. Road building also leads to demands for increasing mineral extraction, places to dump spoil removed from cuttings and motorway service areas. The flood of road associated planning applications, for example to build out-of-town shopping centres and other retail outlets around trunk road junctions, is yet another environmental threat not normally considered when a road scheme is planned.

New and enlarged roads cause damage to protected landscapes through the introduction of alien and discordant features. These may include new road cuttings and embankments, which either scar the landscape or increase traffic noise and road lighting, which in turn degrades their tranquil and remote character.

**WALES**

The *Head on Collision Report (1996)* investigation revealed proposals for road construction which have a primary impact, crossing at least 58.4 km and 55.5 ha of protected wildlife habitat. Many of the schemes identified had no information on the number of kilometres of habitat affected by the roads; therefore the area identified as being under threat will be a conservative estimate. While the secondary impacts can be as devastating to wildlife and habitats as primary impacts, the former is much harder to quantify.

However, these affects are damaging and long term will impact on large areas of sensitive habitats in Wales.

A total of 69 roads schemes were identified which collectively impacted upon a total of 131 important Wildlife Sites. Road schemes impacted on 29 (out of the 131) sites within or adjacent to SSSI's. National Parks, Ramsar sites, AONB and County Parks are being or will be impacted on in some way by road schemes.

A variety of wildlife including badgers, otters, bats, birds, dragonflies, newts and many others are in immediate danger. A wide range of locally, regionally, nationally and internationally important habitats are seen to be under threat.

## SCOTLAND

The result of the 1996 Scottish survey has revealed that 92 schemes will have a direct impact on wildlife habitats. They will dissect a total of 254 km of semi-natural habitat and will affect 101 sites, 17 of which are of either national or international importance. 126 kms of road will cut through woodland, of which 16% is scrub, scattered trees and parkland. As well as the impact on woodland, 65 kms of proposed schemes will affect grasslands, 18 kms heathland, 15 kms peatland, 6 kms raised bog, 3 kms blanket bog and 5 km other bogs and mires. Road building will have a critical impact on the fragile hydrology of many of these ecosystems. A total of 12 kms of road is proposed to go over open water habitats, which are vital for a huge variety of species.

Serious though the damage to designated sites will be, perhaps of greater importance is the damage threatened to semi-natural areas outside designated sites. Scotland has a high proportion of these areas compared with the rest of Britain and they are generally of very high quality. Often areas that have no official designation are of high value to wildlife, visitors, as well as to those who live and work nearby.

It seems clear that some change in the way that we assess environmental, economic and social impacts must be made. Recognition must be given to the value of Scotland's natural heritage as a whole if the damage is to be avoided.

## CONCLUSION

Due to change in the political climate and increasing awareness of the needs of people and their environment there have been a number of changes to the UK Government's transport policy and to the road building programme. This is shown by the reduction from over 300 road schemes affecting over 800 wildlife sites in the pre-1997 roads programme, to 165 road schemes affecting some 300 wildlife sites post the 1997 review. It is worth noting that these 165 road schemes are those under active consideration and not all of the rest have been dropped, so may still pose a threat in the future.

Even with these alterations, the *Head on Collision Reports 1994-1995*, show that there will still be serious direct and indirect impacts on designated and non-designated sites. The effects of habitat severance as a result of a road scheme whether the scale of the scheme is large or small has been clearly recorded and described. The loss of our wildlife habitats is a growing national problem that the road building schemes are continuing to compound.

In addition the quality of the Environmental Impact Assessments (EIA) must continue to be reviewed. The EIA centre in Manchester concluded that between 1988 and 1991 only 25% of the submitted Environmental Statements were considered to be 'satisfactory'. Without adequate assessment it is not surprising that the impact of the schemes is so often damaging.

The need for accurate information is paramount. It is clear that there is little information readily available on the impact of roads and traffic linked to individual species. What there is, is usually being funded by charitable and educational institutions and is reliant on the goodwill of a few committed individuals. Rather than starting new research projects, it may be better to link into existing work such as the Otter and Rivers or Red Squirrel Projects. As well as investment in ecological studies there is a vital need for a co-ordinated mechanism to supply information to policy makers and transport planners. The National Biodiversity Network would answer this need giving ready access to accurate information on habitats and species. At the very least we need well resourced Biological Record Centres able to supply adequate information across the whole country.

Finally, the *Head on Collision Reports* by the Wildlife Trusts of Britain (see Appendix 1) support the growing trend for the debate to be moved from just how to avoid the sites to how to avoid building the roads. Local and National Government and non-Governmental organisations are recognising what many people have already realised. It is far more sustainable to plan for people's needs to be satisfied locally and reduce the need for travel, than to continue short-sightedly adding to an already over-stretched road network.

## ACKNOWLEDGEMENTS

The report has been compiled with the assistance of many individuals and organisations such as WWF-UK, RSPB, Countryside Commission, English Nature, Plant Life, Friends of the Earth, Biodiversity Challenge and the Wildlife Trusts*. We would also like to thank The Mammal Society, Wildcru, The Otter Project and the Vincent Wildlife Trust.

Personally, I thank Carol Hatton (WWF), Barnaby Briggs (RSPB), Ken Robertson (Co Com), plus Tim Sands and Caroline Steele (The UK office of Wildlife Trusts); Martin Newman, Chris Pryor, Alys Langdale and Suzie Holt of the Surrey Wildlife Trust for funding, data inputting and much else.

*      It has borrowed much from the original *Head on Collision Reports (1994-97)* published by the Wildlife Trusts and, in particular, *Head on Collisions in the South East (1994)* by Jonathan Cox.

## REFERENCES

**Byron H, 2000.***A Good Practice Guide for Road Scheme.* Bedfordshire: RSPB (for WWF, RSPB, English Nature and Wildlife Trusts).

**Cox J, 1994.** *Head on Collisions 1994 A report for The Wildlife Trusts in the South East.*

**DETR, 1997.** *What Role for Trunk Roads in England.* London: DETR.

**DETR, 1998a.** *A New Deal for Trunk Roads in England.* London: DETR.

**DETR, 1998b.** *New Approach to Trunk Road Appraisal.* London: DETR.

**DETR, 2000.** *Guidance on Methodology for Multi-Modal Studies.* London: DETR.

**DETR, 2000.** *Transport 2010, The 10 Year Plan.* London: DETR.

**English Nature, 1993.** *Roads and nature conservation, Guidance on impacts, mitigation and enhancement.*

**HA, 1999.** *Highway Agency Environmental Strategic Plan.*

**Mawhinney D, MP, 1995.** *Transport, A Way Ahead.* London: DOT.

**RSNC, 1990.** *Head on Collision Road Building & Wildlife in South-East England.*

**SEEDA, 1999.** *Building a World Class Region, An Economic Strategy for the South East of England.*

**APPENDIX**
**Names of Head On Collisions**

**Table 1.** The Wildlife and Roads Reports for Great Britain 1994-1997

| Names of Head On Collisions | Counties | Date Published |
|---|---|---|
| South East | Surrey/East Sussex/West Sussex/Kent/Hampshire and Isle of Wight | 1994 |
| Yorkshire | Yorkshire and Humberside | 1994 |
| Heart of England | Shropshire/Staffordshire/ Urban (Birmingham)/ Warwickshire and Worcester | 1995 |
| North West | Cumbria/Lancashire/ Merseyside/Cheshire/Greater Manchester | 1995 |
| East Midlands | Derby/Leicestershire and Rutland/Nottinghamshire | 1995 |
| Wales | | 1996 |
| Scotland | | 1996 |
| SouthWest | Cornwall, Devon, Somerset/ Gloucestershire/Dorset/Wiltshire Unitary Authorities: Bristol/Bath and North-east Somerset, South Gloucestershire and North Somerset | 1996 |
| Shires | Bedfordshire/Berkshire/ Buckinghamshire/Cambridgeshire/ Essex/ Hertfordshire/ Northamptonshire/Oxfordshire | 1997 |

**Table 2.** Wildlife Site Impacts due to the English Trunk Road Programme 1994-1997

| COUNTY | SCHEMES | INTERNAT | SSSI | CWS | ASNW | ws | TOTALS |
|---|---|---|---|---|---|---|---|
| Kent | 25 | 2 | 11 | 25 | 9 | 13 | 60 |
| Surrey | 15 | 6 | 5 | 44 | 5 | | 64 |
| W.Sussex | 10 | 0 | 0 | 12 | 5 | 1 | 18 |
| E. Sussex | 11 | 1 | 7 | 9 | 2 | 0 | 19 |
| Hants | 10 | 3 | 7 | 2 | 6 | 12 | 30 |
| I.O.W. | 3 | 0 | 2 | 0 | 0 | 2 | 4 |
| | | | | | | | |
| Cornwall | 11 | 0 | 5 | 12 | 1 | 2 | 20 |
| Devon | 1 | 0 | 0 | 5 | 0 | 3 | 8 |
| S-Set | 6 | 0 | 0 | 16 | 0 | 0 | 16 |
| Brist/Bath | 4 | 0 | 0 | 8 | 0 | 0 | 8 |
| Gloucs | 12 | 0 | 0 | 25 | 0 | 0 | 25 |
| Dorset | 19 | 7 | 11 | 14 | 0 | 2 | 34 |
| Wilts | 12 | 0 | 6 | 0 | 12 | 1 | 19 |
| | | | | | | | |
| Beds | 2 | 0 | 2 | 4 | 0 | 0 | 6 |
| Berks | 4 | 1 | 3 | 11 | 0 | 0 | 15 |
| Bucks | 4 | 0 | 0 | 17 | 0 | 2 | 19 |
| Cambs | 4 | 0 | 3 | 16 | 0 | 0 | 19 |
| Essex | 16 | 0 | 3 | 43 | 0 | 2 | 48 |
| Herts | 4 | 0 | 1 | 4 | 0 | 5 | 10 |
| NHants | 4 | 0 | 1 | 3 | 0 | 0 | 4 |
| Oxon | 4 | 0 | 0 | 4 | 0 | 0 | 4 |
| | | | | | | | |
| Warks | 13 | 0 | 1 | 27 | 0 | 0 | 28 |
| Staffs | 4 | 0 | 1 | 22 | 0 | 0 | 23 |
| Urban | 3 | 0 | 1 | 0 | 0 | 0 | 1 |
| Worcs | 6 | 0 | 5 | 31 | 0 | 0 | 36 |
| H & W | 9 | 0 | 0 | 12 | 0 | 0 | 12 |
| Shrops | 6 | 0 | 2 | 2 | 0 | 2 | 6 |
| East Mids | 40 | 0 | 14 | 17 | 1 | 99 | 131 |
| | | | | | | | |
| Cumbria | 3 | 0 | 0 | 6 | 0 | 0 | 6 |
| Lancs | 9 | 1 | 2 | 27 | 0 | 2 | 32 |
| G Man | 9 | 0 | 0 | 26 | 0 | 1 | 27 |
| Cheshire | 14 | 1 | 1 | 19 | 0 | 1 | 22 |
| Yorkshire | 36 | 0 | 8 | 38 | 0 | 18 | 64 |
| **Totals** | **333** | **22** | **102** | **501** | **41** | **172** | **838** |

| REGION | SCHEMES | INTERNAT | SSI | CWS | ASNW | WS | TOTALS |
|---|---|---|---|---|---|---|---|
| SE | 74 | 12 | 32 | 92 | 27 | 32 | 195 |
| SW | 65 | 7 | 22 | 80 | 13 | 8 | 130 |
| SHIRES | 42 | 1 | 13 | 102 | 0 | 9 | 125 |
| E MIDS | 40 | 0 | 14 | 17 | 1 | 99 | 131 |
| HEART | 41 | 0 | 20 | 94 | 0 | 4 | 118 |
| YORKS | 36 | 0 | 8 | 38 | 0 | 18 | 64 |
| N WEST | 35 | 2 | 3 | 78 | 0 | 14 | 97 |

**Table 3.** HEAD ON COLLISIONS IN WALES 1996
Impacts of the Welsh Roads Programme on Wildlife Sites

| AREA | SCHEMES | SITES | INTERNAT-ONAL SITE | SSI | pSSSI | NAT. PARK | AONB | CWS | WS |
|---|---|---|---|---|---|---|---|---|---|
| North Wales (Clywd / Gwynedd | 23 | 38 | | 6 | 1 | 4 | 1 | 24 | 2 |
| Dyfed (incl. Pembrokeshire) | 16 | 16 | 1 (Ramsar) | 7 | | | | | 9 |
| Montgomery shire | 5 | 5 | | 4 | | | | 1 | |
| Glamorgan | 5 | 24 | | | | | | 24 | |
| Gwent | 2 | 5 | 1 (SAC) | 4 | | | | | |
| Brecknock | 4 | 7 | | 1 | | 1 | | | 5 |
| Radnorshire | 14 | 36 | | 6 | | | | 27 | 5 |
| **Totals** | **69** | **131** | **2** | **28** | **1** | **6** | **2** | **76** | **21** |

The summary form *Head-on Collision 1996*, the Welsh Wildlife Trusts' *Wildlife and Roads Report* shows that 66 road schemes will have a direct impact on 102 Wildlife Sites. Schemes impacted on 29 sites within or adjacent to SSSI, National Parks, Ramsar sites, AONB and Country Parks.

**Table 4.** HEAD ON COLLISIONS IN SCOTLAND 1996.
Impacts of the Scottish Roads Programme on Wildlife Sites.

| AREA | SCHEMES | SITES | INTERNAT-IONAL SITES | SSSI | pSSSI | FNR | AGLV NSA LDA | WC SNCI CWS | WS-ASNW |
|---|---|---|---|---|---|---|---|---|---|
| **Scottish Office** | | | | | | | | | |
| Borders | 2 | 0 | | | | | | | |
| Central | 2 | 3 | | | | | 2 | 1 | |
| Dumfries & Galloway | 3 | 3 | | | | | | | 3 |
| Fife | 2 | 2 | | | 1 | | | 1 | |
| Grampion | 4 | 0 | | | | | | | |
| Highland | 3 | 3 | | 2 | | | 1 | | |
| Lothian | 4 | 10 | 1(pSPA) | | | | 1 | | 4 |
| Strathclyde | 8 | 36 | | 2 | | 1 | 2 | | 4 |
| Tayside | 0 | 0 | | | | | | | |
| Total | 28 | 57 | 1 | 4 | 1 | 1 | 6 | 33 | 11 |
| **Regional Authority** | | | | | | | | | |
| Borders | 4 | 0 | | | | | | | |
| Central | 8 | 2 | | | | | | 2 | |
| Dumfries & Galloway | 3 | 1 | | | | | 1 | | |
| Fife | 0 | 0 | | | | | | | |
| Grampian | 3 | 7 | | | | | | 7 | |
| Highland | 7 | 5 | | 3+1 NNR | 1 | | | | |
| Lothian | 15 | 15 | | 1 | | | | 10 | |
| Strathclyde | 14 | 7 | | 1 | | | 1 | 4 | |
| Tayside | 10 | 7 | | 1 | | | | 6 | |
| Total | 64 | 44 | 0 | 7 | 1 | | 2 | 29 | 5 |
| **Total** | 92 | 101 | 1 | 11 | 2 | 1 | 8 | 62 | 16 |

**Table 5.** Shows that 307 Wildlife Sites were under threat from the 1997 Trunk road Programme.

| REGION | TOTAL SITES EFFECTED | TOTAL SCHEMES | DIRECT IMPACTS | INDIRECT IMPACTS |
|---|---|---|---|---|
| North East | - | 4 | - | - |
| North West | 44 | 22 | 12 | 32 |
| Yorkshire and the Humber | 11 | 15 | 11 | - |
| East Midlands | 71 | 23 | 49 | 22 |
| West Midlands | 43 | 7 | 40 | 3 |
| Eastern | 28 | 26 | 28 | - |
| South West | 27 | 20 | 26 | 1 |
| South East | 69 | 30 | 69 | - |
| London | 14 | 18 | 8 | 6 |
| **Total** | **307** | **165** | **243** | **64** |

**Table 6.** Showing the type of Wildlife Site under threat form the 1997 Trunk Road Programme in England

| DETR Region | INTERNAT-IONAL | NNR | SSSI | LNR | CWS | ASNW | WS | NP | AONB | TOTALS |
|---|---|---|---|---|---|---|---|---|---|---|
| NE | | | | | | | | | | |
| NW | 1 | | 3 | | 23 | 1 | 16 | | | 44 |
| Yorkshire & Humberside | | | 2 | | 8 | | | 1 | | 11 |
| East Mids | | | 10 | | 56 | | 5 | | | 71 |
| West Mids | | | 6 | 1 | 34 | | 2 | | | 43 |
| Eastern | | | 3 | | 25 | | | | | 28 |
| SW | 2 | | 2 | | 19 | | 3 | | | 27 |
| SE | 3 | | 11 | | 36 | 8 | 9 | | 1 | 69 |
| London | | | | 1 | 5 | | 8 | | | 14 |
| **Total** | **6** | **2** | **37** | **2** | **206** | **9** | **43** | **1** | **1** | **307** |

**Table 7.** Impacts of the British Roads Programme on Wildlife Sites

| COUNTRY | ROAD SCHEME | SITES | INTERNAT- TIONAL SITES | SSSI | PSSSI | NAT PARK | AONB | CWS & LNR | WS & ASNW |
|---|---|---|---|---|---|---|---|---|---|
| Wales DTp, LPA | 69 | 131 | 2 | 28 | 1 | 6 | 2 | 74 | 21 |
| Scotland DTp, LPA | 92 | 101 | 1 | 11 | 2 | 1 | 8 | 62 | 16 |
| England DTp | 165 | 307 | 6 | 39 | | 1 | 1 | 208 | 52 |
| **Total** | **326** | **539** | **9** | **78** | **3** | **8** | **9** | **344** | **89** |

## Glossary of Abbreviations

| | |
|---|---|
| TTP | Transport Policies and Programme |
| DTp | Department of Transport |
| SPA | Special Protection Area |
| SSSI | Site of Special Scientific Interest |
| SNCI | Site of Nature Conservation Importance |
| CHS | County Heritage Site |
| ASNW | Ancient Semi-Natural Woodland |
| AONB | Area of Outstanding Natural Beauty |
| WS | Wildlife Site |
| Internat. | Site recognised to be of international significance |
| CWS | County Wildlife Site |
| pSSSI | Proposed SSSI |
| AGLV | Area of Great Landscape Value |
| FNR | Forest Nature Reserve |
| NSA | Nitrate Sensitive Area |
| NNR | National Nature Reserve |
| NP | National Park |
| DETR | Department of Transport, Environment and Regions |

Wildlife Trusts in most administrative areas produced a *Head on Collision Report*. The Wildlife Trusts from each region can be contacted via the National Office.

# 23

# The A36 Salisbury bypass:
# Case Study One

M REED & P WILSON

The proposed A36 Salisbury Bypass was withdrawn from the Government's road-building programme in July 1997. It was originally proposed in 1988 as an 18 km single carriageway with roundabout junctions and this was the basis for the original public consultation and route selection.  By 1991, it had grown into a dual carriageway with grade-separated junctions.  In 1992 the cost was estimated by DoT at £76 million (DoT, 1992).  By 1997, estimates were between £150 million and £200 million (Buchan, 1997).  The route chosen by the DoT is shown in Figure 1. Ecological assessments only took place after the route selection stage, allowing little room for changing the route of the road for ecological reasons.

At the public enquiry, objectors queried traffic forecasts, the economic case for the road, hydrology, ecological and landscape impact, validity of the ecological assessment, pollution figures, the engineering design etc. No statutory bodies were represented. The enquiry inspector's report was published in October 1996 and approved the scheme largely as proposed (Buchanan, 1995). At the same time however, the Government ordered a review of the ecological, hydrological and landscape impacts of the road, to be carried out in consultation with English Nature, The Countryside Commission, The Environment Agency and Wessex Water (Highways Agency, 1997).

The need for a review was prompted by the large amount of evidence that had emerged subsequent to the public enquiry. This included the Standing Advisory Committee on Trunk Road Assessment (SACTRA, 1994) report on traffic induction, The Royal Commission on Environmental Pollution report (1994) on transport and the environment and the MTRU report on the economics of the Salisbury bypass (Buchan, 1997). At the time of the public enquiry there were no statutory conservation designations applying to any part of the route. By 1996 however, the Avon river system had been designated an SSSI and was proposed as a SAC, and the East Harnham Meadows had also been scheduled as an SSSI. The river system

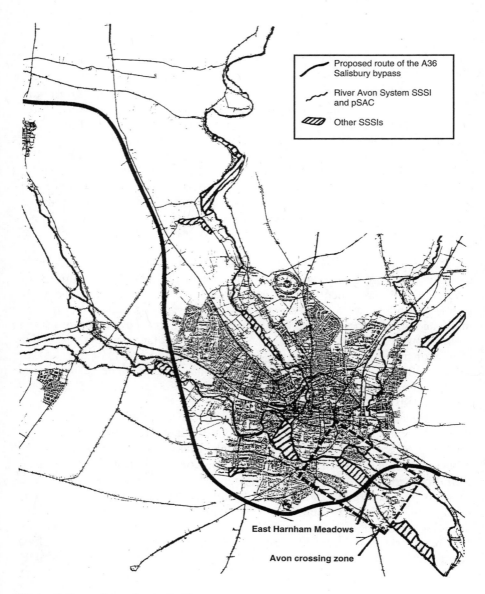

**Figure 1.** Route chosen by the DoT

would have been crossed in six places by the road, and 1.1 ha of the East Harnham Meadows would have been destroyed. These new statutory designations formed the basis of English Nature's response to the review process (Highways Agency, 1997).

The ecological assessment carried out for the DoT concentrated on the three river crossing zones, on the Wylye at Stapleford, on the Nadder between Harnham and Quidhampton, and on the main Avon between East Harnham and Petersfinger. These were seen as the areas of highest potential ecological value. Surveys of these areas were not carried out to English Nature Phase II standard, and vegetation was not assigned to NVC stand types. Part of the Avon crossing was not surveyed at all. No attempt was made to survey other parts of the Avon Valley system in order to establish the importance of any habitats in a local context, and neither were habitats placed in a national context. The area of the East Harnham Meadows which would have been lost was considered of low importance in relation to the rest of the meadow complex (DoT, 1991). The route-corridor away from the river crossings was surveyed superficially only, and a population of the red-data book plant pheasant's eye *Adonis annua* on the line of the road was not located. The Avon Valley system around Salisbury is known to be one of the main strongholds of the barn owl *Tyto alba*, a declining species known to suffer considerable traffic-related mortality (C. Shawyer, *pers. comm.*) yet this species was not included in the DoT's evidence.

In 1994, English Nature carried out a comprehensive survey of habitats adjoining the River Avon and its tributaries in Wiltshire. The aim of this was to determine how much adjacent land should be included within the River Avon System SSSI. The East Harnham Meadows were the largest and richest site discovered during this survey. Unimproved grassland is extremely rare in the Avon Valley system within Wiltshire, and only four other examples were considered of SSSI quality (Highways Agency, 1997). East Harnham Meadows were of such importance that they were given SSSI status in 1995, in advance of the scheduling of the rest of the Avon Valley system in 1996.

The East Harnham Meadows contain approximately 15 ha of the rare MG8 grassland (*Cynosurus cristatus - Caltha palustris* grassland), of which only 500 ha are thought to exist in Britain. This type of grassland may be restricted to Britain. It is typical of river valleys, with a high water table throughout the year, but seasonally inundated with well-oxygenated flowing water (Rodwell, 1992). Grasslands in such situations have frequently been partially improved by conversion to water-meadows, and although this has happened at East Harnham, the unimproved grassland character has been retained. Drier areas in the east of the SSSI have MG5a grassland. The field, which would have been destroyed by the road, was at least as rich as the rest of the site.

East Harnham Meadows are of exceptional species-richness. MG8 grassland has a mean of 26 species per 2 m x 2 m quadrat (Rodwell, 1992), while at East Harnham the mean was 39 species, the richest areas having more than 50 species. The grasslands have no dominant species, but *Festuca rubra, Holcus lanatus, Festulolium* x *loliaceum, Cynosurus cristatus, Carex panicea, C. disticha, Geum rivale, Caltha palustris, Filipendula ulmaria, Juncus articulatus, Ranunculus acris, Trifolium pratense, T. repens, Galium uliginosum, Lotus uliginosus* and *Lychnis flos-cuculi* are abundant. The uncommon *Triglochin palustris* is also frequent, and other unusual species include *Dactylorhiza praetermissa, D. incarnata, Ophioglossum vulgatum, Menyanthes trifoliata* and *Valeriana dioica*. Taller vegetation in less well-grazed areas is closer to the fen community M22b, with abundant *Juncus inflexus, Carex acutiformis, Iris pseudacorus, Pulicaria dysenterica, Thalictrum flavum* and *Glyceria maxima*.

The River Avon and its tributaries contain five species (sea lamprey *Petromyzon marinus*, brook lamprey *Lampetra planeri*, bullhead *Cottus gobio*, Atlantic salmon *Salmo salar* and Desmoulin's whorl snail *Vertigo moulinsiana*) and one habitat (floating vegetation of *Ranunculus* spp.) covered by EU directive 92/43. The Avon has the most diverse fish fauna of any British river, and the invertebrate fauna is also very rich. Populations of Cetti's warbler *Cettia cetti*, reed warbler *Acrocephalus scirpaceus*, sedge warbler *Acrocephalus schoenobaenus*, kingfisher *Alcedo atthis*, sand martin *Riparia riparia*, snipe *Capella gallinago* and redshank *Tringa totanus* are associated with riverine habitats, water voles *Arvicola amphibius* still occur, and signs of otters *Lutra lutra* have been noted in recent years. Aquatic vegetation is dominated by floating mat-forming species including *Ranunculus penicillatus var pseudofluitans, Callitriche* spp., *Potamogeton pectinatus, P. perfoliatus, P. lucens,* with other species including *Sagittaria sagittifolia, Oenanthe fluviatilis, Apium nodiflorum, Butomus umbellatus* and *Oenanthe crocata* (Highways Agency, 1997).

The ecological impact of the proposed Salisbury Bypass as assessed by the DoT differed considerably from the evaluations of English Nature and other bodies carried out during the post-enquiry review (DoT, 1991; Highways Agency, 1997). English Nature's contribution to the review also recommended research required before any road scheme could be approved. This included studies of the hydrology of the Avon river crossing zone and a Phase II survey of all grasslands in the route corridor. It was also recommended that a full research and monitoring programme be conducted into water quality, flow characteristics, physical characteristics and biological features of the Avon and its tributaries (Highways Agency, 1997).

The evidence presented by the DoT in support of the proposed bypass was considered by several statutory bodies as insufficient to evaluate its ecological impact (Highways Agency, 1997). Considerably more detailed evidence is therefore required when assessing new road schemes. In particular, effects on adjacent habitats must be addressed, route corridors must be surveyed thoroughly, surveys must be carried out using acceptable methods and all habitats and species must be considered in local, national and international contexts. Full environmental impact assessments should be carried out at the earliest possible stage, so that environmental constraints can be fully considered when determining the need for any new road scheme.

## ACKNOWLEDGEMENT

We would like to acknowledge the support for this work from the A36 Corridor Alliance and Transport 2000

## REFERENCES

**Buchan K. 1997.** *Transport in Salisbury, Demand Management Solutions.* London: Metropolitan Transport Research Unit.

**Buchanan P. 1995.** *A36 Salisbury Bypass, Concurrent Public Local Inquiries, 20th April 1993-27th April 1994.*

**Department of Transport. 1991.** *A36 Salisbury Bypass, Ecological Survey.* Exeter: Department of Transport, South West Regional Office.

**Department of Transport. 1992.** *A36 Salisbury Bypass Environmental Statement, Non-Technical Summary.* Bristol: Department of Transport, Southwest Construction Programme Division.

**Highways Agency. 1997.** *A36 Salisbury Bypass, Post-Inquiry Review.* Highways Agency, Southern Operations Division.

**Rodwell J. 1992.** *British Plant Communities, 2, Grasslands.* Cambridge: Cambridge University Press.

**Royal Commission on Environmental Pollution. 1994.** *Eighteenth report: Transport and the Environment.*

**SACTRA. 1994.** *Trunk Roads and the Generation of Traffic.* London: HMSO.

# 24

# M40 Shabbington habitat creation area: Case Study Two

## C J BICKMORE

The construction of the M40 (Waterstock to Wendlebury section to the east of Oxford) resulted in an incursion into the periphery of Shabbington Wood Site of Special Scientific Interest (SSSI) situated between the M40 and Shabbington Wood on the Oxfordshire/ Buckinghamshire borders. In this locality the motorway alignment was agreed on the condition that an area of over 4 ha was given over, planted and managed for nature conservation purposes on behalf of the Highways Agency (Bickmore, 1992). The allocated land was adjacent to Shabbington Wood and comprised a section of field in arable production and one in a grass ley (Figure 1a,b). The SSSI was designated on account of the entomological interest

## DESIGN OBJECTIVES AND IMPLEMENTATION

The aim of habitat creation work was to attract a number of species of butterfly present in the locality in particular the rare black hairstreak butterfly *Strymonidia pruni*, together with other locally occurring species including the brown hairstreak *Thecla betulae*, white letter hairstreak *S. w-album* and purple emperor *Apatura iris*. The design aimed to provide a sheltered habitat using plant species attractive to these butterflies. Site preparation began in 1990 with the planting of a series of irregular bands of trees and shrubs, particularly blackthorn *Prunus spinosa*, and the sowing of a local hay seed crop. The plantings and the existing hedgerows/woodland edge were managed to encourage butterflies. A 25 year management plan has been prepared also for the site.

## MONITORING

The development of the grassland was monitored using a random sample of quadrats. The presence of brown and black hairstreak eggs was

**LEGEND**

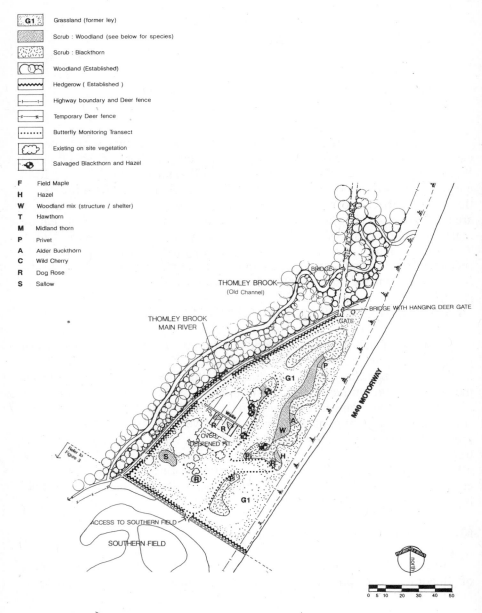

| | |
|---|---|
| G1 | Grassland (former ley) |
| | Scrub : Woodland (see below for species) |
| | Scrub : Blackthorn |
| | Woodland (Established) |
| | Hedgerow ( Established ) |
| | Highway boundary and Deer fence |
| | Temporary Deer fence |
| | Butterfly Monitoring Transect |
| | Existing on site vegetation |
| | Salvaged Blackthorn and Hazel |

| | |
|---|---|
| **F** | Field Maple |
| **H** | Hazel |
| **W** | Woodland mix (structure / shelter) |
| **T** | Hawthorn |
| **M** | Midland thorn |
| **P** | Privet |
| **A** | Alder Buckthorn |
| **C** | Wild Cherry |
| **R** | Dog Rose |
| **S** | Sallow |

**Figure 1a.** Site plan of northern field.

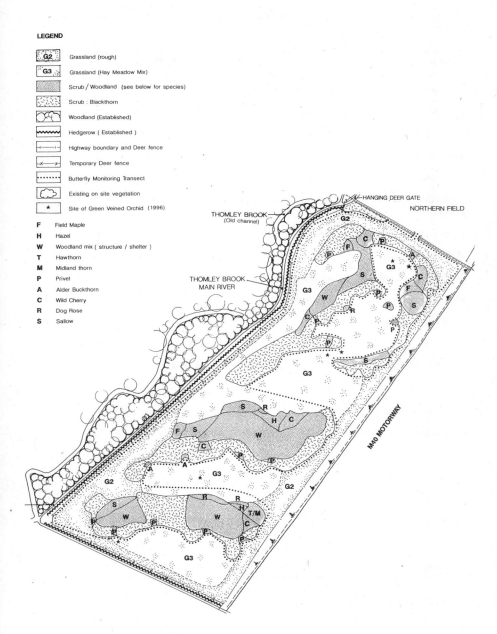

**LEGEND**

| | |
|---|---|
| G2 | Grassland (rough) |
| G3 | Grassland (Hay Meadow Mix) |
| | Scrub / Woodland (see below for species) |
| | Scrub : Blackthorn |
| | Woodland (Established) |
| | Hedgerow ( Established ) |
| | Highway boundary and Deer fence |
| | Temporary Deer fence |
| | Butterfly Monitoring Transect |
| | Existing on site vegetation |
| ★ | Site of Green Veined Orchid (1996) |

| | |
|---|---|
| **F** | Field Maple |
| **H** | Hazel |
| **W** | Woodland mix ( structure / shelter ) |
| **T** | Hawthorn |
| **M** | Midland thorn |
| **P** | Privet |
| **A** | Alder Buckthorn |
| **C** | Wild Cherry |
| **R** | Dog Rose |
| **S** | Sallow |

**Figure 1b.** Site plan of southern field.

**Table 1.** Changes in the annual index of abundance of adult butterflies recorded along BMS transects in the M40 conservation area.

| Species | Vernacular Name | 1991 | 1992 | 1993 | 1994 |
|---------|-----------------|------|------|------|------|
| *Thecla betulae* | Brown hairstreak | 1 | 0 | 0 | 1 |
| *Gonepteryx rhamni* | Brimstone | 14 | 18 | 24 | (1) |
| *Aricia agestis* | Brown argus | 0 | 0 | 0 | 5 |
| *Polygonia c-aldum* | Comma | 10 | 8 | 2 | (1) |
| *Polyommatus icarus* | Common blue | 7 | 67$^{++}$ | 102$^{+++}$ | 31$^{+}$ |
| *Thymelicus flavus/ T. lineola* | Essex / Small skipper | 141 | 153 | 286$^{+++}$ | 473$^{+++}$ |
| *Pyronia tithonus* | Hedge brown | 211 | 172$^{-}$ | 184 | 142 |
| *Ochlodes venatus* | Large skipper | 21 | 66 | 57 | 69$^{+}$ |
| *Melanargia galathea* | Marbled white | 1 | 1 | 1 | 7 |
| *Maniola jurtina* | Meadow brown | 130 | 277$^{+++}$ | 300$^{+++}$ | 268$^{+++}$ |
| *Anthocharis cardamines* | Orange tip | 23 | (6) | (16) | (4) |
| *Inachis io* | Peacock | 92 | 152 | 78 | (19) |
| *Vanessa atalanta* | Red admiral | 4 | 2 | 2 | 0 |
| *Aphantopus hyperantus* | Ringlet | 83 | 65 | 135 | 126$^{+}$ |
| *Lycaena phlaeas* | Small copper | 14 | 8 | 3 | 4 |
| *Coenonympha pamphilus* | Small heath | 14 | 36 | 10 | 5 |
| *Aglasis urticae* | Small tortoiseshell | 60 | 92 | (6) | (0) |
| *Artogeia rapae* | Small white | 39 | 44 | 60$^{++}$ | 24 |
| *Celastrina argiolus* | Holly blue | 10 | 0 | 1 | 0 |
| *Cynthia cardui* | Painted lady | 1 | 4 | 0 | 2 |
| *Pararge aegeria* | Speckled wood | 9 | 34 | 30 | 37 |
| *Limenitis camilla* | White admiral | 0 | 0 | 2 | 0 |

Significant increases compared to national trends for species: $^{+}$P<0.05, $^{++}$P<0.01, $^{+++}$P<0.001; decrease $^{-}$P<0.01. () indicates underestimate due to missed transect counts

**Figure 2.** Existing woodland edge deer-fencing has enabled a more irregular profile to develop and blackthorn to sucker into the northern field.

recorded by examining blackthorn for their eggs. Other butterfly species were monitored by the local members of Butterfly Conversation using the method of the national Butterfly Monitoring Scheme (Pollard & Yates, 1993), and recordings taken along a fixed transect (Table 1). This enabled comparisons with elsewhere in the country.

## FINDINGS

Findings are presented from the first five years of monitoring. By 1995, the planted trees and shrubs were well established and the blackthorn was becoming suitable for black hairstreak. Along the existing woodland edge and hedges an irregular profile had developed and provided sheltered pockets. The species composition of the sown grassland changed from one where *Holcus lanatus* and *Trifolium pratense* were strongly dominant to one where more interesting species were becoming established including *Orchis morio*, *Primula veris* and *Lathyrus nissolia*. Some of the changes are attributed to changes in nutrient levels (Bickmore & Thomas, 1999).

**Table 2.** Changes in actual brown hairstreak egg numbers recorded in different parts of the M40 conservation area

| Area | 1991 | 1992 | 1993 | 1994 |
|---|---|---|---|---|
| Wood edge (plot A) | 23 | 10 | 10 | Not searched |
| Suckers along wood edge (plot A) | 18 | 80 | 27 | 23 |
| Wood edge (plot B, E) | 9 | 20 | 18 | Not searched |
| Hedge between two fields | 3 | 6 | 32 | 8 |
| Hedge on west boundary | 0 | 0 | 2 | 0 |
| Planted bushes | 20/250 | Not searched | 51/100 | 21/100 |
| (No. eggs / no. plants examined) | | | | |

Brown hairstreak egg counts (Table 2) showed that densities greatly creased in the habitat creation area making it one of the larger colonies in Britain. The decline in 1994 reflected the poor season for the species everywhere. Genuine colonies of marbled white and brown argus first appeared in 1994. The white-letter hairstreak was first recorded in 1995. With improved shelter, black hairstreak is expected soon.

## POSTSCRIPT

Monitoring of both grasslands' development and butterfly populations has continued. The species richness of the grassland has increased with seven more species being recorded in 1998 than 1991. Desirable species including *Orchis morio* and *Lathyrus nissola* persist.

In 1991 black hairstreak were first recorded. The brown hairstreak population has remained at previous levels. The findings show that it is possible to restore valued communities of wildlife to heavy, intensively farmed lowland agricultural land. Monitoring and habitat management both are important in this respect. In this instance through focusing on specific species land associated with highway construction has been developed to increase biodiversity.

## REFERENCES

Bickmore CJ. 1992. M40 Waterstock-Wendlebury: planning, protection and provision for wildlife. *Proceedings of the Institute of Civil Engineering.* **June 93:** 75-83.

Bickmore CJ, Thomas JA. 1999. The development of habitat for butterflies over land in former arable cultivation. *Aspects of Applied Biology* **58:** 305-312.

Pollard E, Yates T. 1993. *Monitoring butterflies for ecology and conservation.* London: Chapman and Hall.

farmed lowland agricultural land. Monitoring and habitat management both are important in this respect. In this instance through focusing on specific species land associated with highway construction has been developed to increase biodiversity.

## REFERENCES

**Bickmore CJ. 1992.** M40 Waterstock-Wendlebury: planning, protection and provision for wildlife. *Proceedings of the Institute of Civil Engineering.* **June 93:** 75-83.

**Bickmore CJ, Thomas JA. 1999.** The development of habitat for butterflies over land in former arable cultivation. *Aspects of Applied Biology* **58:** 305-312.

**Pollard E, Yates T. 1993.** *Monitoring butterflies for ecology and conservation.* London: Chapman and Hall.